WOLFRAM
SUMMER SCHOOL
RESEARCH REPORTS
2024

Edited by Mohammad Bahrami

Wolfram Summer School Research Reports 2024

Edited by Mohammad Bahrami
Copyright © 2026 by Wolfram Media, Inc.

Wolfram Media, Inc.
wolfram-media.com
ISBN 978-1-57955-109-4 (paperback)
ISBN 978-1-57955-110-0 (ebook)

For information about permission to reproduce selections from this book, write to permissions@wolfram.com.

Typeset with Wolfram Notebooks: wolfram.com/notebooks

First edition.

Table of Contents

Preface

MOHAMMAD BAHRAMI

Director of Academic Innovation Support

The Wolfram Summer School remains one of the most distinctive educational environments in the world, not only for the work produced but also for the model of learning it embodies. As generative artificial intelligence accelerates the shift toward experiential and heuristic forms of learning, the Summer School has long operated at this frontier. What follows is a snapshot of the projects from the 2024 program, but just as important is the culture and collaborative ecosystem that make these projects possible.

At the heart of the Summer School is our commitment to experience-driven, action-oriented learning. Students do not merely study concepts; they build, test, revise and ship ideas into the world. This aligns with a broader transformation in education: moving beyond front-loaded knowledge and simulated academic exercises toward learning environments where students gain fluency through iteration, feedback and authentic problem solving.

A Community of Diverse Thinkers and Practitioners

Each year, our cohort brings together participants from a wide array of backgrounds, geographically, academically and professionally. In 2024, students arrived from dozens of countries and represented a spectrum of experience: from undergraduates to industry professionals, from physicists and computer scientists to designers, mathematicians and entrepreneurs.

Our mentors share an equally diverse mix of expertise. Many are seasoned Wolfram developers, while others are longtime collaborators who bring deep knowledge from across scientific, technical and creative fields. The continuity of mentorship, where former students become mentors, is one of the most powerful indicators of our program's impact.

This diversity is not incidental—it is foundational. It creates the rich environment in which students learn heuristics, adopt new modes of reasoning and build the personal and professional networks that support innovation well beyond the program.

Mentorship as a High-Contact, Collaborative Process

Learning at the Summer School is intentionally high contact. Each student works closely with a small team of mentors who guide the development of their project. But the school's ethos is rooted in the belief that "it takes a village," an idea that mirrors the collaborative nature of contemporary research and industry workflows.

Students benefit from an environment where experts across Wolfram Research can be brought into a conversation at a moment's notice, whether at a scheduled meeting, a lunchtime discussion or a late-night debate on theory, strategy or design. Through this, students see not just what Wolfram Research builds, but *how* it builds: through iterative problem solving, open exchange and intellectual curiosity.

One of the most remarkable aspects of this experience is the direct and enthusiastic involvement of Stephen Wolfram himself. He engages passionately with every student—at meals, on walks, in corridors and in impromptu conversations, bringing an unmatched depth of insight and an infectious excitement for ideas. He actively challenges mentors to help students make real, substantive progress on their projects, pushing the boundaries of what they can achieve in just a few weeks.

For mentors, this level of engagement can be intense: long meetings, deep technical discussions and sustained focus. But the results—seeing students grow, create and accomplish far beyond their initial expectations—make the experience profoundly rewarding every single year.

Project-Based Learning, Reimagined

The most tangible outcome of the Summer School is the project each student produces. But these projects are not academic exercises. They are real contributions, prototypes, tools, algorithms, experiments, analyses and models that extend well beyond the bounds of a classroom assignment.

The 2024 projects span a remarkable range of inquiry, reflecting the interdisciplinary nature of both Wolfram Language and contemporary scientific and technological challenges:

- Large language models (LLMs): from identifying personality traits to building LLM agents for data science, RAG systems for arXiv data, machine learning pipeline generators and democratic participation simulations

- Physics, cosmology and fundamental models: investigations into black hole mergers, discrete spacetime gravitational radiation, rewriting-rule–based physics, gyrogeometry and multiway systems

- Computation, algorithms and theory: exploring call graphs, Turing machines, hypergraphs, automata evolution, minimal axiom systems, tensor network optimization and extensions of the Mandelbrot set

- Chemistry, materials and applied sciences: spanning electrolyte modeling, chemical space networks, halite fluid-inclusion analysis, quantum chemistry using quantum computing and high-school natural-law education

- Engineering, robotics and interfaces: from robotic surgery coordinate systems to robotic arm control using LLMs, power system analysis, renewable energy models and mind-controlled BCI keyboards

- Data, society and systems thinking: including social media prototypes, Medicaid reimbursement analysis, invasion sports modeling, Moon crater selection for telescopes and problem-solving ecosystems

This breadth showcases how experiential learning unleashes creativity: when students confront real problems with real tools, they produce ideas and artifacts that are both rigorous and deeply original.

A Model for the Future of Education

The Wolfram Summer School embodies what we as members of the Academic Innovation Support team at Wolfram believe education must become: a collaborative ecosystem where students, faculty and industry practitioners operate as coinvestigators; where learning is inseparable from doing; where heuristics, workflows and real-world decision making stand alongside conceptual knowledge.

For over two decades, the Summer School has served as our laboratory for this model. Now, through expanded partnerships and collaborations with universities, professional groups and organizations, we're working to make this type of learning accessible at scale—embedded throughout academic programs rather than limited to elite institutions or isolated experiences.

Looking Ahead

The projects highlighted here are only a fraction of the intellectual activity that took place during the 2024 Summer School, yet they demonstrate the shared themes that continue to define our program: curiosity, rigor, creative computation and the drive to contribute something meaningful.

We remain grateful to every student, mentor and collaborator who makes this environment possible. The work that begins here often extends far beyond our time together, into research careers, entrepreneurial ventures and ongoing collaborations across our growing global community.

PROJECT LIST

The Project List is a catalog of abstracts for all student projects from the Wolfram Summer School 2024. Scan the QR code or visit the URL on each page to view the full project and interact with the code.

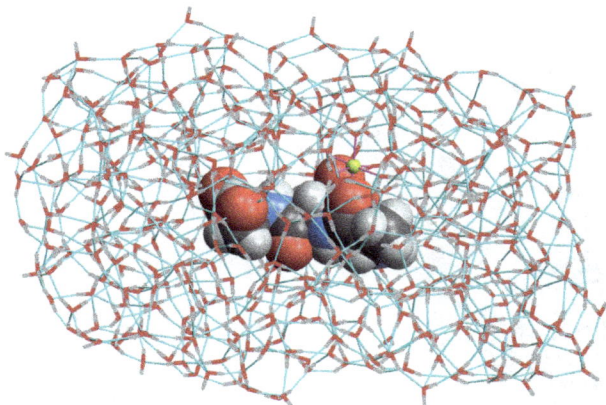

Modeling of Interactions in an Aqueous Electrolyte with a Peptide Additive

XUELIAN LIU

To clarify the positive effect of a peptide additive in an aqueous electrolyte, a model has been built by using the MoleculeComplex package for molecule manipulation, energy optimization and interaction visualization for the first time in Wolfram Language. An initial test of the intermolecular energy capability of the Merck molecular force field (MMFF) has been performed and the density of water nanoballs has been calculated for further verification of model reliability. In an optimized water nanocylinder molecule complex with a tripeptide at the center and a cation nearby, we can observe hydrogen bonds in turquoise blue between water molecules, and between water molecules and the peptide as well. In addition, the first solvation shell of the cation coordinates water molecules at five octahedral sites and the cation coordinates oxygen in the carboxyl group in the peptide at the sixth octahedral site.

Scan or visit
wolfr.am/WSS2024–Liu

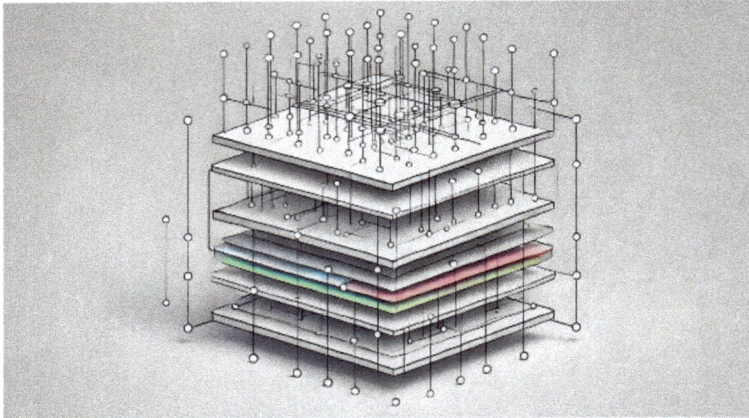

Identifying and Manipulating Personality Traits in Large Language Models

RUMI ALLBERT

The field of large language models (LLMs) has experienced rapid growth in recent years, driven by the pursuit of improved performance, interpretability and safe utilization. This project builds upon work in "activation engineering," such as "Refusal in LLMs Is Mediated by a Single Direction" and "Steering GPT-2-XL by Adding an Activation Vector," to explore the realm of personality manipulation in LLMs. We propose a novel approach to identify and manipulate personality-related activation directions, potentially enabling dynamic fine-tuning of LLM personalities. This project simultaneously aims to enhance our understanding of LLM interpretability while addressing the ethical implications of such advancements.

Scan or visit
wolfr.am/WSS2024–Allbert

Particles in a Box: Kinetic Gas Theory and Simulations

JAKUB TRZASKA

The aim of this project is to give intuitive insight into the elusive nature of a gas contained in a box through the use of the kinematic approach. We investigate the dynamics of a 1D hard-sphere system, also referred to as a hard-rod gas. By employing the event-driven simulation method, we compute the evolution of a system of colliding particles and visualize the properties of the model. We model a container with two species of gas particles separated by a wall and explore the dependency on relevant parameters.

Scan or visit
wolfr.am/WSS2024–Trzaska

Search for Rules that Preserve Geometric Structure in Wolfram Models

FERGUS NOBLE

The Wolfram model is a class of models consisting of a graph (or hypergraph) and a rewriting rule. It has been shown by [1] and [2] that the Wolfram model can exhibit behavior corresponding to general relativity in the continuum limit; however, it remains unknown precisely which rules show this correspondence. In this project, I conducted a systematic search over the space of possible rewriting rules to identify rules that preserve the geometric structure of graphs that correspond to static vacuum solutions of the Einstein equations, a necessary condition for these rules to satisfy general relativity. I considered rules involving two and three edges on simple directed graphs corresponding to flat Minkowski and Schwarzschild solutions. From a search space of 49,116 rules, I identified 32 that preserve geometric structure and may be candidates for modeling general relativistic behavior.

Scan or visit
wolfr.am/WSS2024–Noble

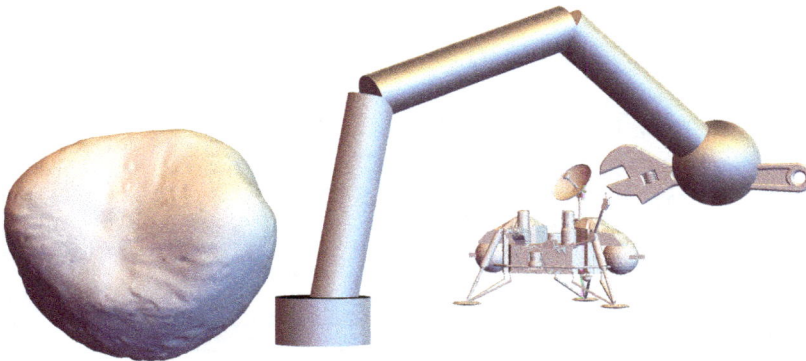

Robotic Arm Control Using Large Language Models

CURRAN FLANDERS

The use of large language models (LLMs) can add natural language comprehension and reasoning abilities to a robot control system. This project uses the capabilities of LLMs to enable a robotic arm to follow natural language instructions. The arm is simulated in Mathematica in an environment containing other objects. First, a 2D model of the arm is constructed. Then the arm is extended into 3D. Functionality is added to allow the arm to be used in different environments. Each environment contains a collection of 3D objects. Finally, functionality is added to allow the LLM to guide the movement of the arm by specifying a list of sequential target positions.

Scan or visit
wolfr.am/WSS2024–Flanders

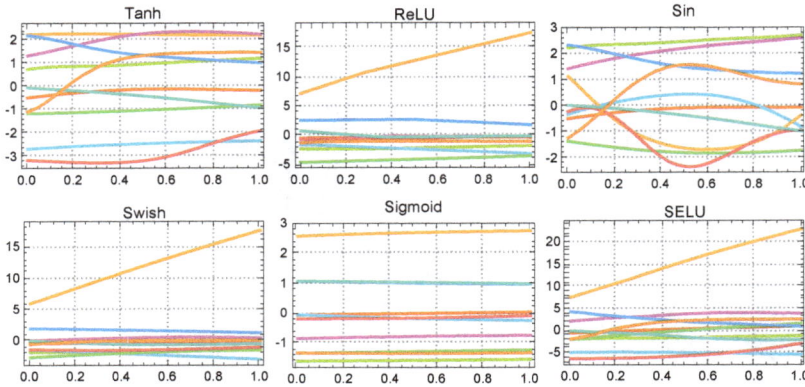

Exploring the Distributions of Shallow Multi-layer Perceptrons

TEO BENAROUS

This work explores various statistical and mathematical techniques for analyzing and representing the distributions of shallow multi-layer perceptrons with two hidden layers and a width of five. The fixed architecture adheres to the universal approximation property as proven by [2]. To assess the potential of common activation functions, we have conducted a descriptive and comparative analysis of their functional spaces. We start by examining the basic statistics and additionally provide empirical estimates of the first four moments. A significant portion of the study focuses on the Karhunen–Loève expansion (KLE) and its application in deriving principal functional components from empirical covariance operator estimates. We analyze the approximation errors from truncated KLEs and investigate the learned distributions for smooth activations, highlighting typical functions and anomalies. The discrete wavelet transform (DWT) is employed to explore the spaces of networks induced by the non-smooth activation functions. We discuss coarse coefficients, the energy fraction of details and the main wavelet coefficients' projection to a lower-dimensional manifold. Additionally, we assess the learned distributions within this context.

Scan or visit
wolfr.am/WSS2024–Benarous

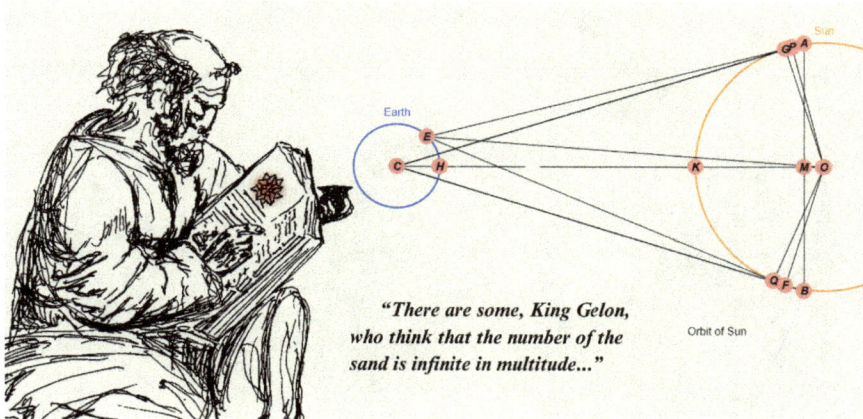

"There are some, King Gelon, who think that the number of the sand is infinite in multitude..."

The Sand Reckoner, Now in Wolfram Language

SHAN YANG YAU

This project aims to recreate Archimedes's Psammites (The Sand Reckoner) using Wolfram Language, making it more accessible to a wider audience by means of Archimedes's friendly tone and interactive diagrams. The goal is to imbue the reader with both a sense of satisfaction in understanding the text and a sense of appreciation of Archimedes's incredible work, all while assuming minimal mathematical knowledge. There is still some additional work to be done to complete a full representation of The Sand Reckoner, which will be pursued after the Wolfram Summer School.

Scan or visit
wolfr.am/WSS2024–Yau

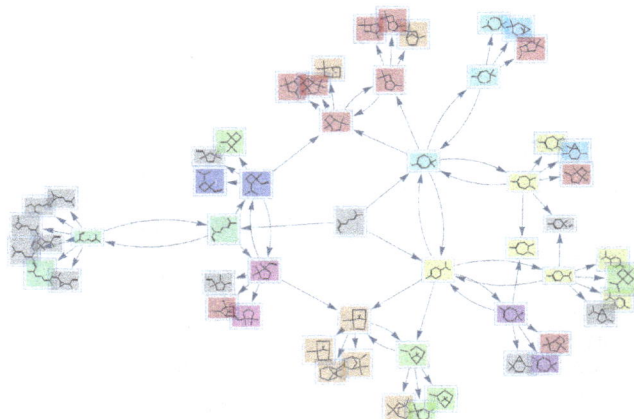

Analyzing Multiway Chemical Reaction Networks Using Chemical Space Networks

NICHOLAS FRIELER

This project combines multiway chemical reaction networks with chemical space networks to analyze structural similarities between the products of different reaction pathways with identical or closely related precursor molecules. It explores several different avenues for analysis that can be paired with these methods, including the number of clusters in chemical space networks, the size of the largest such cluster and branchial graph analysis. It also verifies the validity of chemical space network clustering using multiway genealogical distance and principal component analysis on the molecules produced by the multiway chemical reaction network.

Scan or visit
wolfr.am/WSS2024–Frieler

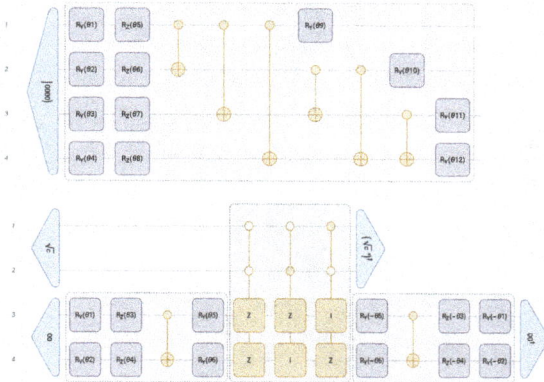

Electronic Structure of Atoms Using Quantum Computing Techniques

MICHAŁ ZDZIENNICKI

This project aims to simulate the ground state of hydrogen-like atoms using quantum computing tools in Mathematica. It employs quantum circuits to represent fermionic operators, following the principles of the Hartree–Fock theory. By applying mixed quantum-classical algorithms like the variational quantum eigensolver, the goal is to minimize energy within the circuits, thereby transforming initial states into the ground state of a Hartree–Fock Hamiltonian. This research demonstrates the potential of quantum computing in simulating atomic structures and computational chemistry.

Scan or visit
wolfr.am/WSS2024–Zdziennicki

Building Blocks' Aggregation Systems

MARIA FERNANDA CASTRO ALVAREZ

Aggregation systems on grid graphs grow at each step by attaching a new atom to a preexisting configuration according to a set of local constraints. The multiway graphs of these systems fascinate us by sometimes filling space and other times reaching frustrated and dead surfaces. This project continues to explore aggregation systems on three-dimensional grid graphs by introducing unique and indivisible atoms, which adhere to specific combination rules. We examine configurations formed with a few special building blocks, whose interlocking mechanisms allow for practical experiments. These shapes, designed for the physical world, also have an abstract mathematical representation in terms of adjacency relations that form noteworthy graph structures. We introduce the concept of instruction graphs and ensure their uniqueness with regard to constructions in the physical world. By comparing computational models to physical sculptures, we validate the practical applicability of our new approach, highlighting its potential to enhance our understanding and control of space-filling structures.

Scan or visit
wolfr.am/WSS2024—Alvarez

Encoder-decoder transformer model

Design an RNA sequence that folds into a given RNA secondary structure.

```
AGCUCUGAG
AACUGAAUU
CCAUGGGUU
AUAUCAAUG
UCAGACCUG
UGAAAUUCA
GUUCUUCAG
CU
```

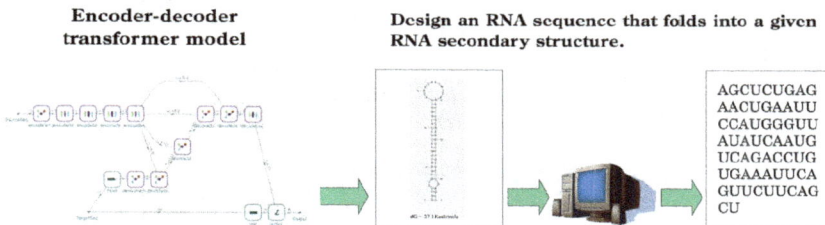

A Language Model for RNA Design in Wolfram Language

DANNY BARASH

Inverse RNA folding, known as RNA design, has been developed since the early 1990s in a variety of directions and has been applied to diverse applications in biotechnology and RNA research. There is also a citizen science game called eteRNA that introduces RNA design and RNA folding principles to human players, with no previous knowledge required. The approach to solving RNA design problems and eteRNA challenges has mostly been centered on a variety of Monte Carlo strategies, but without advanced machine learning except for some data taken from human eteRNA players. Recently, a language model has been proposed to learn a prior on nonhuman data taken from the RNA families database (Rfam) with RNA folding predictions performed, and thereby to improve the performance of a Monte Carlo search strategy. Further, we apply this new approach using Wolfram Language toward the solution of difficult eteRNA challenges and especially the solution of biologically meaningful RNA design problems by improving any Monte Carlo strategy being used, offering initial RNA sequences that are trained by our model instead of random guesses or human-led priors before the mutations start in any Monte Carlo scheme. The language model used for the proof of concept is an encoder-decoder transformer neural network architecture. All three components (the Monte Carlo strategy, the language model and the data) can be further extended for an improved RNA design. In this contribution, we focus on extending the language model and the data.

Scan or visit
wolfr.am/WSS2024–Barash

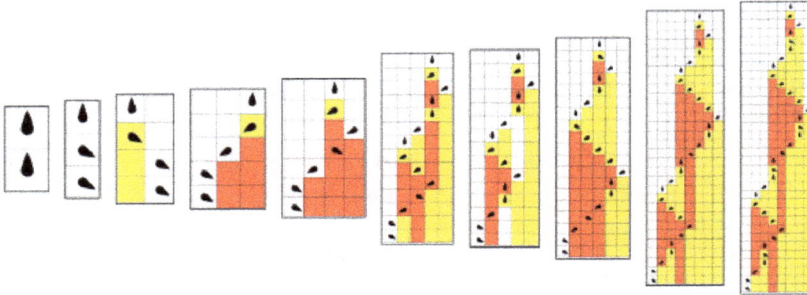

Adaptive Evolution for Turing Machines

BRIAN A. MBOYA

When we think of biological systems, we don't always have a way of describing them and therefore lack an understanding of what's going on underneath these systems as they undergo evolution. This project tries to ignore some of the constraints that we may encounter in biological systems, namely "we don't know what's going on," and essentially think of biological organisms as computational systems with certain sets of rules they follow to develop. These rules are equivalent to the genotype of the organism and the result of executing them is its phenotype. We take a minimal model for biology that has this setup, Turing machines, and study its adaptive evolution.

The main idea is to have a Turing machine with configurations of n-colors, s-states and a rule. We then run the machine for a number of steps, changing the rule at each step to find a pattern that lives the longest.

This is an attempt to solve the famous busy beaver problem, but with a different approach of trying to evolve busy beavers to find one that lives the longest.

Scan or visit
wolfr.am/WSS2024–Mboya

Reversibility in Mobile Automata

ADAM FRENCH

In this work, we study mobile automata, discovering a theorem that fully characterizes their reversibility. We outline the proof and verify these results through computations. We also provide background on how the proof was discovered. By examining their behaviors, we identified reversible mobile automata exhibiting seemingly unpredictable behavior. When restricted to a finite cyclic tape, we found that all reversible automata form closed cycles.

Scan or visit
wolfr.am/WSS2024–French

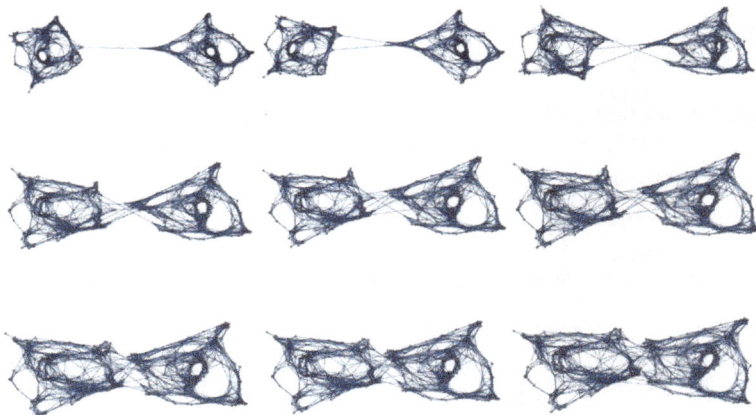

Investigating Binary Black Hole Merger from Rewriting Rules in Wolfram Models

MEPHY LIU

By solving the Einstein evolution equations with Brill–Lindquist initial conditions, binary black hole merger occurs naturally on a spatial graph. However, this method demands substantial computational power, limiting its reproducibility and accessibility. This project aims to explore the potential of simulating binary black hole mergers using rewriting rules in Wolfram models. The approach seeks to provide a more efficient method for studying the merger process, including the evolution of the metric, the generation of gravitational waves and changes in the dimensional structure of spacetime.

Scan or visit
wolfr.am/WSS2024–M–Liu

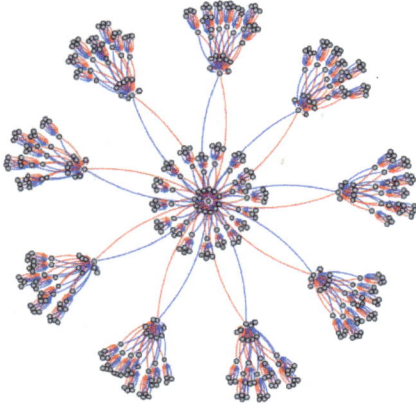

Exploring Call Graphs of Nestedly Recursive Functions

THOMAS ADLER

Nestedly recursive functions are functions that have to call themselves (recursion) multiple times (nesting). Complex patterns emerge from these seemingly simple but nested functional forms. This project first implements a stack-based recursive function from scratch to have more introspection into how the recursion works. Then it explores the function space of various kinds of shift operators. The call graphs of these relatively basic shifting operations exhibit interesting, complex and sometimes fractal patterns. We also explore simple arithmetic operations, although that function space is more limited. We find patterns that are tree- and branch-like, as well as flowers and even a type of marine coral.

Scan or visit
wolfr.am/WSS2024–Adler

Finding Minimal Axioms for Ring Theory

TATE ALLEN

A ring is an algebraic structure defined by a longstanding set of seven axioms. Given that it is often desirable to find the smallest possible set of axioms for different algebraic theories, this project aims to find the smallest possible axiomatic system for the theory of rings through the computational power of Wolfram Language. It was determined by A. Tarski that ring theory may be reduced to a single axiom. A notable candidate for this axiom has been found, but remains unverified. Future work includes verifying the single-axiom system for rings through proof and establishing upper and lower bounds on the lengths.

Scan or visit
wolfr.am/WSS2024–Allen

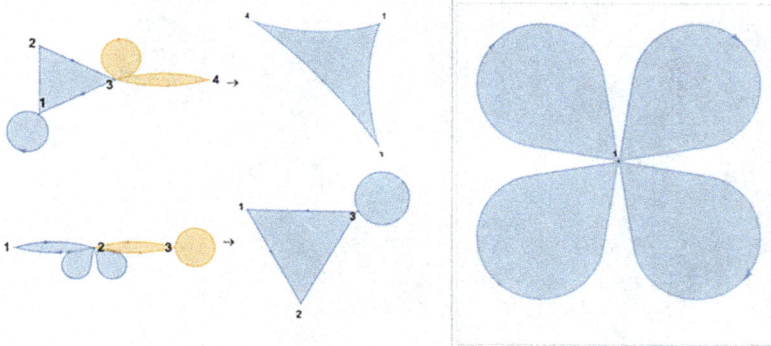

Compositional Paths on Hypergraphs

SERGIO QUIROGA

The notion of sequentiality has always been central in mathematics and at the heart of computation, encoded in objects like the successor function in natural numbers or, as we will study, in paths on graphs. A path on a graph is usually defined as a concatenation of edges, but can also be studied as a chain of compositions of edges, thinking of the edges as binary relations.

We will construct a generalization of compositions of binary relations to compositions of higher-arity relations. Then we will study how chains of compositions of n,m-arity relations applied to a set of hyperedges within a given hypergraph can be thought of as a new kind of path on hypergraphs, called a compositional path.

The goal of this project is to develop a formalism for studying the compositionality of hyperedges and, therefore, paths on hypergraphs. We will compute examples and multiway systems of compositional paths as well as propose new types of path algebras on hypergraphs.

Scan or visit
wolfr.am/WSS2024–Quiroga

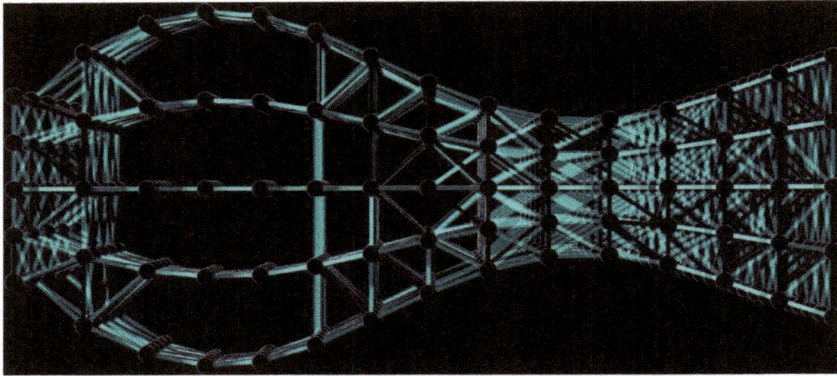

Gravitational Radiation in Discrete Spacetimes

NATHAN MAGARI

This study investigates the radiative behavior of discrete spacetime structures within the framework of the Wolfram Physics Project class of models. Different hypergraph-rewriting rules are used to evolve an initially perturbed discrete spacelike hypersurface and then analyzed to determine how the spatial curvature and graph distance metric evolves, similar to how gravitational waves are studied in general relativity by perturbing the spacetime metric tensor. By analyzing the discrete analog to gravitational waves across different Wolfram models, particularly in scenarios involving compact objects like neutron stars and black holes, this research aims to uncover potential deviations that could expand our understanding of spacetime structure at fundamental scales.

Scan or visit
wolfr.am/WSS2024–Magari

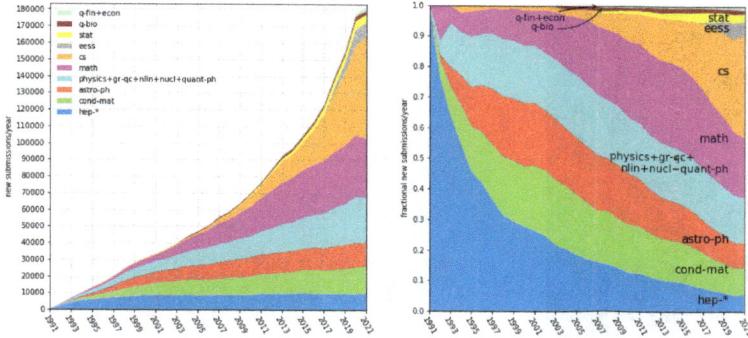

Building a RAG for arXiv Data

MICHAEL ERNEST

The arXiv repository is a vast and growing collection of academic papers that is now capturing a growing audience in industry, in particular for large language models (LLMs) and AI. It's also inspired several side projects from Kaggle and similar platforms aimed at making it easier to access.

It is a daunting corpus of documents to sift through wholesale. The URL identifiers, for one, use an indexing scheme (e.g. https://arxiv.org/abs/1234.56789) that has changed with growth and new categorization. The kinds of papers that compile or restate other research can have whole paragraphs dedicated to meticulous citation. And, of course, there are simply a lot of things. The charts above show just the rate of growth in submissions from 1991, the year arXiv was formed, through 2021. arXiv users could benefit even from integrating summaries of these documents with an LLM. The natural language interaction offered by OpenAI, Claude or similar services makes it plausible to review arXiv content without first learning how best to search it and have the result cast in a conversational manner.

This project is an experiment in that direction. It uses a method of supplementing an LLM with additional data called retrieval-augmented generation (RAG). A number of paper abstracts are collected and included in a prompt sent to the LLM. The user agent asks the LLM to incorporate this data into its response. The end goal is to incorporate this scheme into a Chat Notebook, such that a user agent can prompt the LLM for information that, unaided, it might otherwise respond to in tangential or even inaccurate ways. (It is, after all, the goal of a generative AI system to keep a conversation going, even at the risk of regressing to standard search or calling up suspect data from the corpus.)

Scan or visit
wolfr.am/WSS2024–Ernest

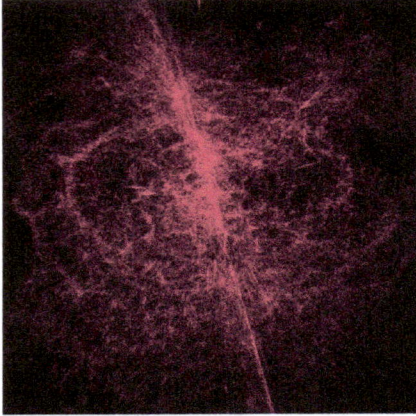

Searching for Indications of Fractional Dimensions

LUKE WRIGLESWORTH

The power law scaling of galaxy distributions sourced from HyperLEDA are measured using growing spherical shells. Measurements are performed in isolation as well as in a three-dimensional grid configuration using hundreds of shells. It is found that the power law exponent is not uniform in isolation, but will always cluster around a particular value when repeated enough times using a sphere radius that is not too large. The results found are unsurprising, and typically center around the expected values of $N(r) \propto r^3$ and $\dfrac{dN(r)}{dr} \propto r^2$, but exceed these values for $r > 1.2 \times 10^9$ light years. The explanation for such a deviation can likely be explained by the introduction of large-scale structures in the spheres causing a lack of uniformity in the distributions.

Scan or visit
wolfr.am/WSS2024–Wriglesworth

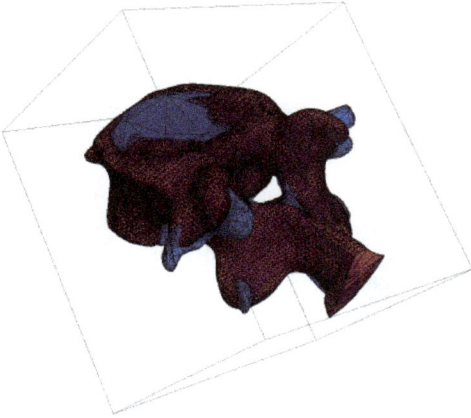

Creating an Anatomic Coordinate System for Robotic Surgery with L4 Vertebra Models

LOGAN NYE

Surgical robots are increasingly utilized to perform surgical procedures in the spine due to their capabilities in making precise, controlled movements. This project focuses on aligning anatomical models using radial basis function (RBF) interpolation to create a precise coordinate system essential for surgical robotics. We detail the process of loading and normalizing both idealized and real patient-specific models of the fourth lumbar vertebra (L4). By visualizing and selecting corresponding anatomical landmarks on these models, we perform RBF interpolation to accurately align the idealized model with the real one. This alignment is critical in developing a standardized coordinate system that enables surgical robots to navigate complex anatomical structures with high precision, enhancing surgical accuracy and consistency. The project aims to advance minimally invasive surgery by integrating robust, data-driven coordinate systems into robotic surgical workflows.

Scan or visit
wolfr.am/WSS2024–Nye

Spot a Cat: Cellular Automata Edition, or Representational Images in Cellular Automata

KATHRYN CRAMER

Stephen Wolfram has tasked me with finding images of cats in cellular automata (CA) based on his essay "Generative AI Space and the Mental Imagery of Alien Minds," in which he explored the computational space around AI-generated images of cats in party hats.

How might we discover the cats in CA? How might we identify representational images in CA by representing words and CA-generated images as vectors? How might we refine these representations?

By applying a reflection to grids generated by 2D CA rules, we generate emoji-style representations of words, which we call Rorschachs in honor of the inkblot test, using vector representations of the images to find the best match for words. This results in representational CA images that both computers and humans can recognize. What we seem to get when we access CA cat space to make Rorschach cats is a rudimentary version of a search tool providing transparency to the CLIP and OpenCLIP Multi-domain Feature Extractors. Delivering access to cat space seems to give a view into the concept space. What we have is a toy model of a tool at a low resolution. The possibilities are very intriguing.

Scan or visit
wolfr.am/WSS2024–Cramer

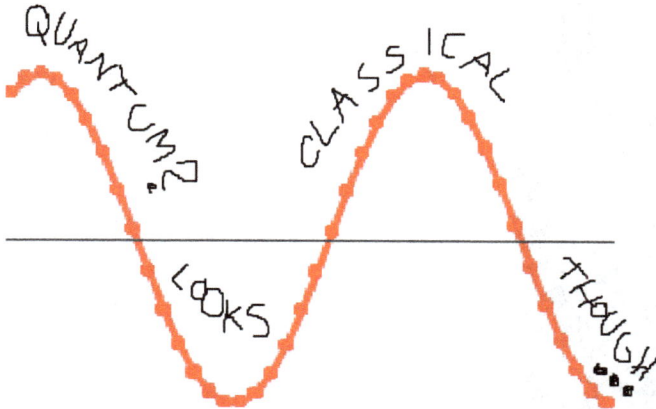

Truncated Fock Space as a Way to Model 2nd Quantization in Wolfram Language

JURAJ TKÁČ

In this project, we were exploring truncated Fock space as a way to model second quantization using Wolfram Quantum Framework. We went through studying states in this basis, in particular coherent and cat states, and visualized their properties with the purpose of providing a good insight into their meaning. That can lead to the discovery of a larger area of quantum computation based on cat qubits, which look to be promising natural error-correcting frameworks toward universal quantum computation.

Scan or visit
wolfr.am/WSS2024–Tkac

A Mind-Controlled Keyboard: Utilizing SSVEPs to Create a Storytelling BCI

EMMA SUSI

Steady-state, visually evoked potentials are electroencephalographic (brain wave) responses to visual stimuli flashing at different frequencies. These potentials allow us to "mind read" and determine which stimuli the participant is looking at by analyzing these responses. This project involves creating a storytelling interface with a digital four-key keyboard that types as a user "selects" a word with their mind. It leverages a large language model (LLM) to display the four next-most probable words live on the page for the user to select.

Type with your mind, and choose your own adventure.

Scan or visit
wolfr.am/WSS2024–Susi

Exploring LLMs' Participation in a Representative Democracy

DERWIN A. PRITCHETT

As large language models (LLMs) continue to proliferate, the contrast between their writing and that of their human counterparts continues to sharpen our focus on what it means to produce meaningful speech. In this project, we explore two proxies an LLM could play if allowed to participate in a democratic election: that of a candidate and of a voter.

The goal is to evaluate the performance of LLMs in these roles by comparing the analyses of their responses to their real-world counterparts and then independently assessing the biases of LLMs using data anonymization, multiple-choice permutations and persona variations. LLMs will be utilized in two distinct scenarios:

1. Analyzing presidential political debates
2. Examining biases in LLM-based assessments of presidential political debates

In this project, we seek to evaluate how well LLMs draw conclusions from political candidates' responses to journalistic prompts developed for a presidential debate and explore any biases, whether implicit or explicit, in their responses.

Scan or visit
wolfr.am/WSS2024–Pritchett

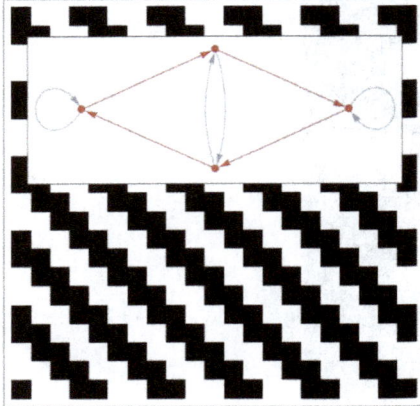

Finding and Bounding Generalized Ising Model Ground-State Energies

DAVID MOORE

Generalized Ising models form a rich area in which one can explore questions of computability and phase transition [1]. While numerical methods to approximate ground states are readily available, few sources enumerate and prove the existence of periodic ground states in the thermodynamic limit. In this project, we follow the branch-and-bound paradigm for bounding the ground state from above [4], and we employ linear optimization with appropriate constraints for bounding the ground state from below [2]. When these bounds are equal, we have provably found a ground state in the thermodynamic limit. By employing general methods—especially those described in Huang, et al. (2016)—this can be done for a surprisingly general class of generalized Ising models.

Scan or visit
wolfr.am/WSS2024–Moore

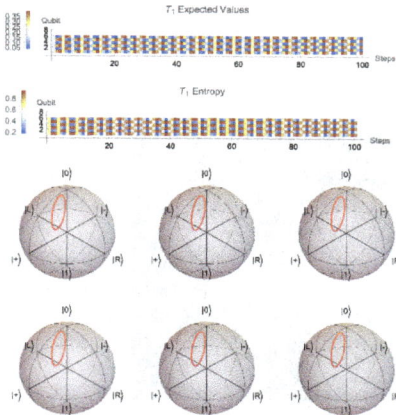

Finding Patterns in Quantum Information: A Computational Approach

ANDRES ARCIA-LOPEZ

Quantum information systems are challenging to visualise, since the dimension of the space grows exponentially with the number of qubits. However, by extracting information about a quantum state, we can more easily visualise meaningful qualities of the system. This project focuses on the iterative flow of information through cellular automata–like quantum circuits.

Scan or visit
wolfr.am/WSS2024–Arcia–Lopez

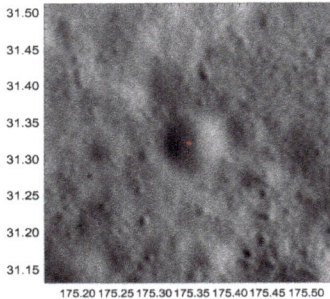

Identifying Candidate Moon Craters for Radio Telescope Placement

ANDREW HERRERA

This project will survey the surface of the Moon and select craters that best approximate paraboloids as candidate locations for the construction of radio telescopes. The project will make use of Moon crater entity positions as well as geo elevation data. By fitting paraboloids to the topography of Moon craters and selecting the best-fitting and most appropriate sites to minimize excavation, materials and work required in construction, as well as solar radiation interference, we will narrow down the list to one candidate for the location of a radio telescope on the Moon.

Scan or visit
wolfr.am/WSS2024–Herrera

A Computational Model for Invasion Sports

WILLEM NIELSEN

Soccer, basketball and hockey are widely practiced and spectated. Understanding the dynamics of these sports is important for players and coaches at all levels. There are obvious similarities between these so-called "invasion sports," and I was curious if I could capture them in a computational model.

Scan or visit
wolfr.am/WSS2024–Nielsen

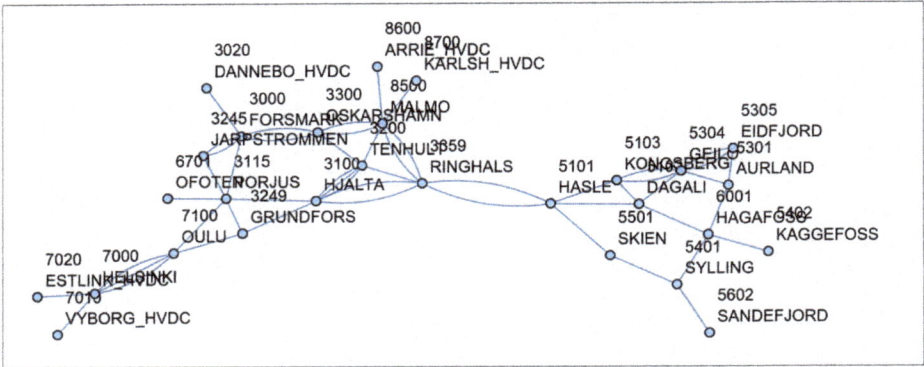

Ad-hoc Electric Power System Analysis with PSSE RAW Files

GLEN HALLEY

Power systems are evolving; however, the availability of efficient collaborative engineering design and analysis tools lags behind the technological deployment in the field. This project leverages the complete design model of the PSSE RAW file and a future extension to the .dyr file format and Common Grid Model Exchange Specifications (CGMES) Common Information Model (CIM) to facilitate the exchange of operational and grid-planning data among transmission operators. Once translated, the depth of Wolfram Language and Wolfram System Modeler combined with OpenIPSL can be used to analyze the data.

Scan or visit
wolfr.am/WSS2024–Halley

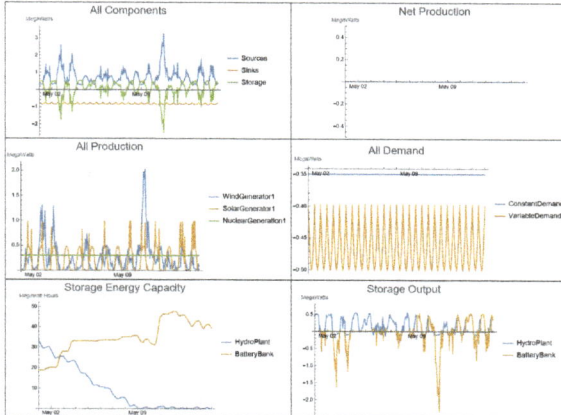

Renewable Energy System Modelling

NIILO KANTONIEMI

This project develops a framework for modelling energy systems. It is designed to accommo-date a wide range of energy production, consumption and storage systems. The framework's flexibility allows it to integrate various types of energy sources, including renewable and nonrenewable resources. By using real-world data, the model can simulate energy system behaviours in any location and from any time range.

Scan or visit
wolfr.am/WSS2024–Kantoniemi

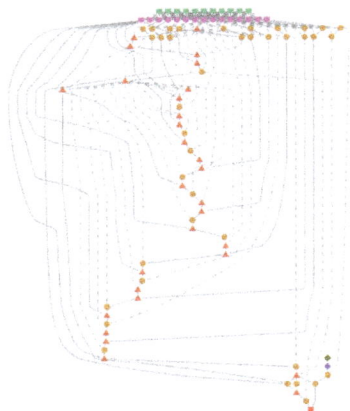

Study "Logic Puzzle" Axiom/Proof Graphs from TPTP.org

ARIHANT GADGADE

There is a repository of theorem-proving problems called Thousands of Problems for Theorem Provers (TPTP.org). This project has the goal of representing a problem in terms of its axiomatic system and conjecture in Wolfram Language and using the automated theorem-proving capabilities of the language to build out proof graphs of the proof to each conjecture. Using these proof graphs will lay the foundations for future analysis of metamathematical properties of the proof space.

Scan or visit
wolfr.am/WSS2024–Gadgade

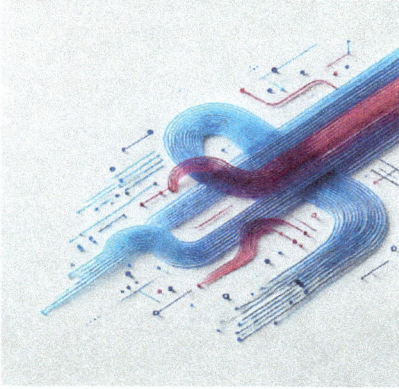

"Composable Intelligence for Science" – GPT-4o

Building LLM Agents for Data Science

ASTITVA CHOPRA

This project explores the integration of large language models (LLMs) with the Wolfram Language stack to build semi-autonomous agents capable of performing diverse data science tasks. Leveraging Wolfram Language's well-curated and computable knowledgebase and functional programming approach, this project demonstrates the use of LLMFunction *and* LLMTool *as potential building blocks of agent-like architectures. This work presents practical examples and future explorations toward developing autonomous workflow agents that generate Wolfram Language code for AI and data science applications. This could be a great tool to increase access to AI in the scientific community by simplifying data preparation and computability.*

Scan or visit
wolfr.am/WSS2024–Chopra

Non-possible Blocks of a Cellular Automaton

ATIK SANTELLÁN

This project investigates the relationship between cellular automata and regular languages. The primary objective is to determine outputs that cannot occur from any input in a given cellular automaton characterized by the number of states (k) and neighborhood radius (r). This can be done by constructing and analyzing de Bruijn graphs. Each graph will be annotated with a particular cellular automaton's transition rules, either with a Moore or Mealy machine structure. The machines can be transformed into a deterministic form where the question of missing inverses reduces to path finding on a directed graph.

Scan or visit
wolfr.am/WSS2024–Santellan

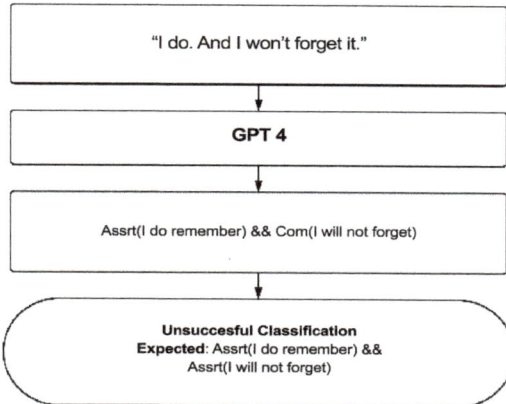

A New Type of Social Media

CARL FLYGT

We have attempted to construct a basis for machine–human ethics using John R. Searle's 1976 analysis of speech acts in natural language. We have demonstrated a large language model (LLM) system capable of rigid classification of speech behavior in conversation. Searle's theory appears to us an overlooked innovation toward a general formal framework for natural language conversation.

Scan or visit
wolfr.am/WSS2024–Flygt

Expansion of the Fractal Mandelbrot Set to 3D Using Dimensional Folding

DAVE MOUTON

An approach using dimensional folding generated an interesting three-dimensional analog of the Mandelbrot set, demonstrating the viability of a three-dimensional analog of the two-dimensional space of complex numbers.

If this exercise represents a canonical 3D analog of the two-dimensional space of complex numbers, exact closed-form solutions to 3D stress, fluid, heat transfer and electromagnetic field problems may be possible. If exact closed-form solutions are possible, exact optimization of the geometry of structural components, airfoils, cooling fins, antennas, etc. may be possible using the potential of symbolic computing vs. iterative methods that are currently employed.

Scan or visit
wolfr.am/WSS2024–Mouton

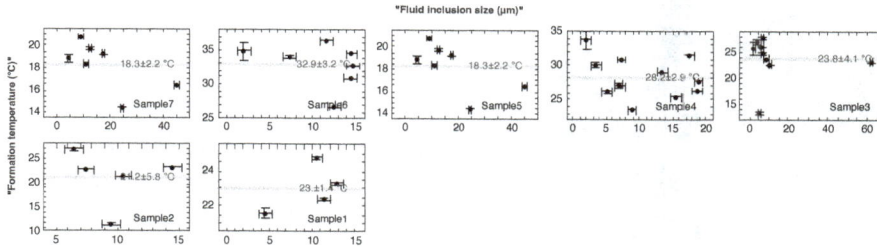

Automated Experimental Data Processing: Example of Halite Fluid Inclusion Paleo-temperatures

EMMANUEL GUILLERM

We develop a code that automates the formatting of input data, the calculation of corrections and the generation of a notebook containing an output dataset and various graphs necessary for halite fluid inclusion paleo-thermometry. The input is a series of spreadsheets (.csv) containing the information of each fluid inclusion (T_{obs}, size, composition, water column height) for each sample. The computational result is a standardized data structure including the input data and the analysis results.

Scan or visit
wolfr.am/WSS2024–Guillerm

Pre-Lie Structures of Graph Rewriting and Associated Axiom Systems

FERNANDO OLIVIE MÉNDEZ MÉNDEZ

The aim of this project is twofold: First, to study the noncommutative algebras that arise in the context of graph rewriting (both context-free and context-sensitive cases). It is illustrated that in certain cases, we always get a pre-Lie structure. Second, to study the axiom systems associated with these algebraic structures—that is, relations of the form $(x \circ y) \circ z - x \circ (y \circ z) = (y \circ x) \circ z - y \circ (x \circ z)$ are considered.

Scan or visit
wolfr.am/WSS2024–Mendez

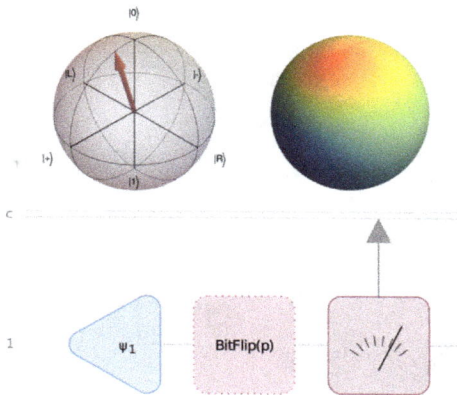

Studying the Effects of Noise in Simple Quantum Logical Systems

STANISŁAW RAKOWSKI

This project investigates the influence of quantum noise on the performance of quantum algorithms. Simple quantum noise models are explored theoretically in both single-qubit and multi-qubit systems. First, we will simulate a one-qubit quantum circuit with a noise channel, studying how the nature and magnitude of the noise can be characterized and estimated using only measurement counts. The results of these simulations will be applied to measurements performed on quantum processing units (QPUs) to characterize the effects of quantum noise in real quantum computing hardware.

Scan or visit
wolfr.am/WSS2024–Rakowski

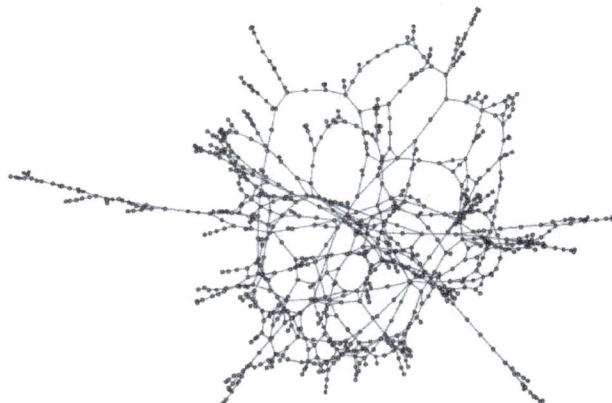

Graph Theory Approach to Hard-Sphere Fluids

GABRIEL MALAVÉ

Hard spheres have been widely used in condensed matter as models for a variety of systems, such as liquids, crystals and glasses. While there exists extensive theoretical insight as well as numerical data, much work is required for characterizing dense amorphous packings (glassy states). In this project, we build upon current efforts to compute the observables of these systems by means of token-event graphs that summarize their dynamical evolution. We investigate the structure of these causal graphs and explore the possibility of drawing a parallel with the Feynman diagram construction to better understand the physical properties reflected in them.

Scan or visit
wolfr.am/WSS2024–Malave

Computational Thinking Education through Natural Law for High-School Students

HE YU

Introduce high-school students to computational thinking, which includes problem analysis, decomposition, abstraction and algorithmic thinking.

Foster an understanding of natural laws such as competition, cooperation and adaptation.

Develop skills in problem solving, critical thinking, teamwork and resilience.

Scan or visit
wolfr.am/WSS2024–Yu

The Equivalence Principle, Hyperbolic Geometry and Gyrogeometry

IBRAHIM BUKVIC

In this project, we aim to understand what happens to gravitation in accelerated frames undergoing a Thomas precession. We first construct causal graphs in flat spacetime and then in asymptotically flat spacetime while investigating their foliations. Once a relationship is established between gyrogeometry and gyrovector spaces with the equivalence principle in this discrete framework, it will have far-reaching consequences in our understanding of concepts such as spacetime heat.

Scan or visit
wolfr.am/WSS2024–Bukvic

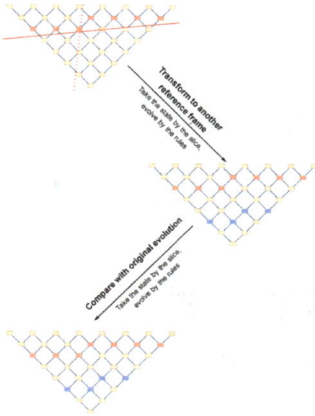

On Motion and Special Relativity in Wolfram Models

LEV CHIZHOV

The goal of this project is to introduce the class of "relativistically coherent" models—the ones that can viably be transformed from one reference frame to another. An example of the implementation of moving particles will be introduced, and special relativity will be examined in the models in question.

Scan or visit
wolfr.am/WSS2024–Chizhov

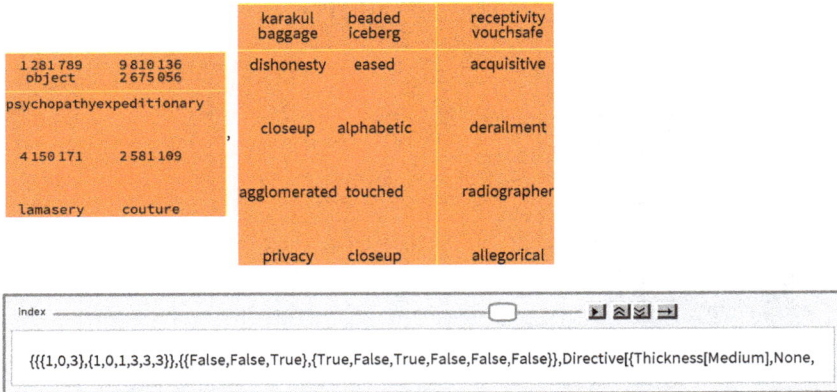

karakul baggage	beaded iceberg	receptivity vouchsafe
dishonesty	eased	acquisitive
closeup	alphabetic	derailment
agglomerated	touched	radiographer
privacy	closeup	allegorical

```
1 281 789     9 810 136
  object      2 675 056
psychopathyexpeditionary

4 150 171     2 581 109


lamasery      couture
```

index ————————————————————————————————————○——— ▶| ≋| ≋| ⇥|

{{{1,0,3},{1,0,1,3,3,3}},{{False,False,True},{True,False,True,False,False,False}},Directive[{Thickness[Medium],None,

Deduce Grid Styles from an Image of a Table Using Neural Networks

LUCAS GREENWOOD

In this project, we train a neural network on a dataset of grid raster images generated with the Grid function. The images are labeled with the style options (converted to strings) that produced them, and the network must learn how to associate images to these style options. In this way, the task can be viewed as a mix between image captioning and a narrow-scoped code-generation task. The final system utilizes a transformer-based architecture to extract the style options, although other architectures—such as convolutional neural networks and recurrent neural networks—were experimented with for demonstrative purposes, as this was my first time working with neural nets. Further work may include the use of text recognition to reproduce cell contents (including spanning entries), word-based embeddings and expansion of the network's utility on real-world images.

Scan or visit
wolfr.am/WSS2024–Greenwood

A Promise Made Is a Debt Unpaid

OLGA COOPERMAN

The way money is created and managed can have significant impacts on economic growth and financial stability. Understanding how money works is important for making informed decisions about personal finance, investing and participating in the economy.

Scan or visit
wolfr.am/WSS2024–Cooperman

Ramanujan sent a letter on infinite series.

You've Got LLMail

PRABHCHINTAN RANDHAWA

TL;DR: LLMs summarize email well.

Ada Lovelace wrote to Charles Babbage in 1840 [1] showing interest in discrete mathematics. Srinivasa Ramanujan wrote to G. H. Hardy in 1913 [2] on the distribution of prime numbers. Much about written communication has changed since—so much so that in 2024, with an exponential volume compared to handwritten letters, (e)mail has an increasingly rising chance of going unnoticed. Would Ramanujan hear back from G. H. Hardy today? With large language models (LLMs) quickly beginning to take over linguistically inclined tasks and in the process generate seemingly infinite words, one starts to wonder how much of it is being generated using LLMs… but more importantly, which communication is worth paying attention to?

The broader inquiry originated while trying to find ways to optimize startup and general-business operations, probing whether a "single-person-company" is possible via modern-day AI-enabled automatons. As a starting point, this project focuses on optimizing email traffic for a given user using Mathematica's MBOX and MailRecieverFunction to: a) analyze one's current inbox for activity and behavioral insights; and b) set up LLM-based simple rules to equip the user with extremely short summary functionalities for a quick glance.

Scan or visit
wolfr.am/WSS2024−Randhawa

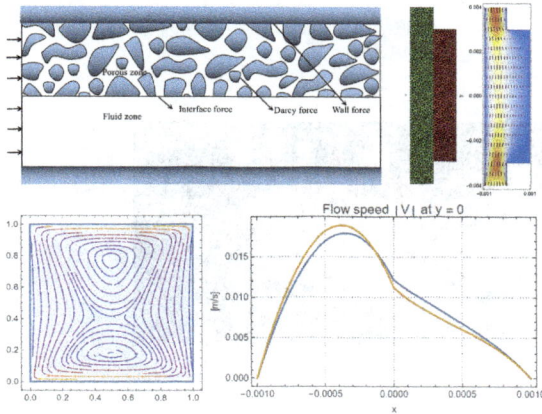

Flow speed |V| at y = 0

Solving Navier–Stokes and Brinkman–Forchheimer PDEs for Coupled Free and Porous Media Flow

SAFI AHMED

This project focuses on the development of computational models for a coupled-flow scenario comprising free and porous media. The fluid flow in the free region is governed by the Navier–Stokes PDE, while the flow in the porous region is governed by the Brinkman–Forchheimer PDE. A numerical solution for the coupled flow has been developed using the finite element method (FEM) framework in Wolfram Language, with conditional statements employed to enforce region-specific PDEs. An implicit issue with the continuity equation in the existing FluidFlowPDEComponent *function was identified and resolved. In the latter part of the project, a custom function,* BrinkmanPDEComponent, *was developed in Wolfram Language to compute flow through porous media. The newly developed function supports both dimensional and nondimensional forms of the Brinkman PDE. Detailed examples, scope and applications of this new function are provided at the end of the post.*

Scan or visit
wolfr.am/WSS2024–Ahmed

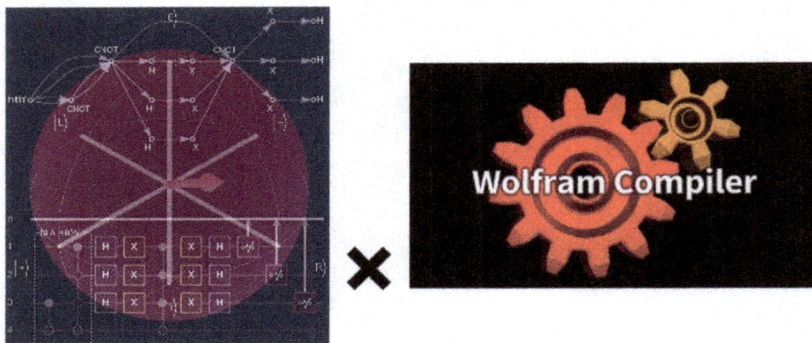

Optimizing Tensor Network Contraction Using the Wolfram Compiler

VEDANT TEWARI

The QuantumFramework paclet integrates quantum computation capabilities into Wolfram Language, providing symbolic representations for quantum bases, states and operators to model and simulate quantum circuits and algorithms. Currently, we are focused on optimizing tensor contraction operations using Einstein summation, which involves updating tensor indices and mapping superscripts to subscripts. The existing implementation, reliant on functionalities like EinsteinSummation, Transpose *and* FindPermutation, *constructs and models network graphs but does not leverage the Wolfram Compiler, leading to inefficiencies. We plan to develop compiled versions of these operations to optimize for speed and memory.*

Scan or visit
wolfr.am/WSS2024–Tewari

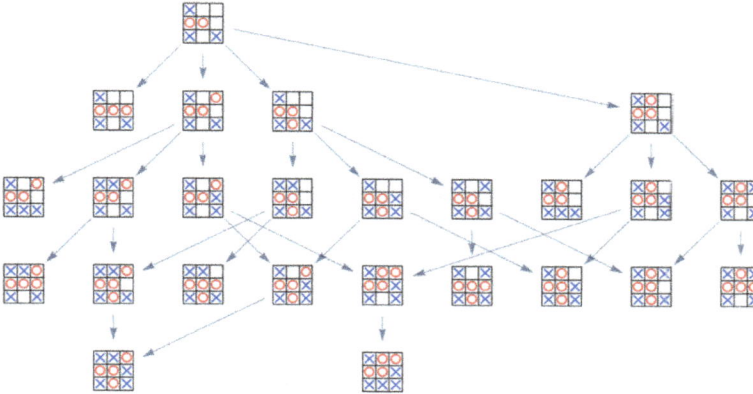

Introducing Combinatorial Games in Wolfram Language's Entity Framework with OOP

ZSOMBOR MÉDER

Combinatorial games are two-player, perfect-information games in which players take actions sequentially. Well-known examples include Tic-Tac-Toe, Connect 4 and Nim. This project aims to create a suitable representation structure using Wolfram Language's entity framework and to demonstrate the framework's viability. Ensuring compatibility with existing Wolfram Language functions is of particular importance.

Scan or visit
wolfr.am/WSS2024–Meder

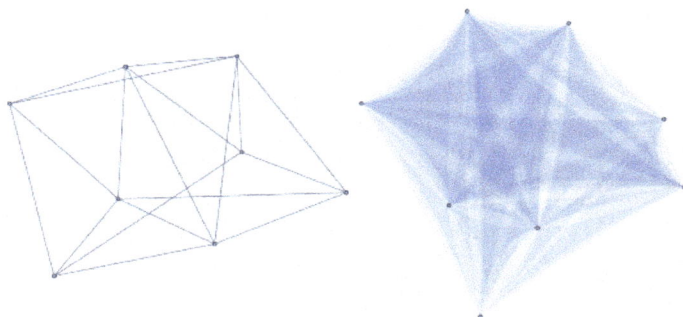

Morphology of Hypergraphs: A Study of Hyperedge Evolution

THOMAS MURFF

This project develops a theoretical technique for "morphing" the hyperedge sets of hyper-graphs, in accordance with particular symmetries among the hyperedges. The goal is to explore a notion of hypergraph morphology. In terms of the theory, by encoding an arbitrary hypergraph into a Boolean hypercube via a Boolean function, we can perform group-theoretic analyses on the hypercube. These group symmetries are permutation isometries on the hypercube, where the intuition is to consider k-arity hyperedges as analogous to k-arity faces of shapes in geometry, which guides us to consider symmetries that respect analogous (but not precisely) the same properties. This algebraic information can be used to update the hyperedge set to a new one, which is defined via a second encoding Boolean function. A given pair of hyperedges will either ascend or descend in arity to a new encoded hyperedge that either covers them or is covered by them, when they are connected by a permutation isometry. In this project, there are two main algorithmic functions at work that rely on refining or eroding the hyperedges, the former moving up in arity and the latter moving down. The theory, as well as the implementations in Wolfram Language, can allow one to derive an effective morphology of hypergraphs via smoothing or refining the hyperedge set. One can then look toward the various data structures that hypergraphs may represent, and then infer what such a morphology would mean in that context. Moreover, this project defines code that visualizes the "space" of all applications of these two functions, in the form of a multiway evolution system. These visualizations suggest that the symmetries among the hyperedges of a hypergraph are highly related to this space of all transformations of a hypergraph, and conversely that asymmetric and "rough" hypergraphs will have convoluted spaces of transformations.

Scan or visit
wolfr.am/WSS2024–Murff

GC001 fit with {24h, 12h} cosinor

TRef:1989-12-22T00:00:00

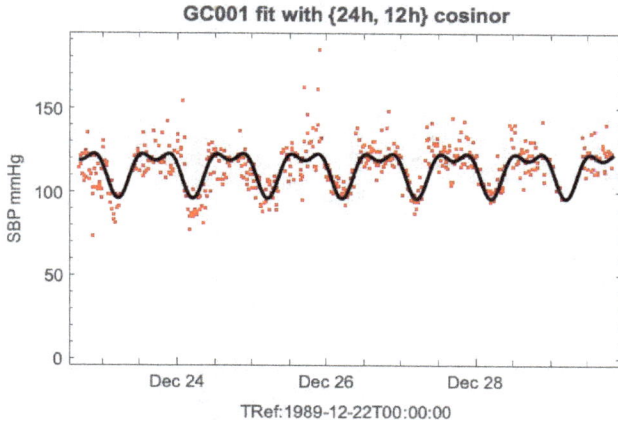

Implementing Cosinor Rhythmometry in Wolfram Language

CHASE TURNER

The single cosinor method [1,2,3,4] is employed to model the circadian variation in physiological variables. While the 24-hour component usually accounts for the largest proportion of variability in the data, harmonic terms contribute to shape the circadian rhythms. CosinorFit is introduced both for single and multicomponent cosinor model fitting of measured physiological variables.

Scan or visit
wolfr.am/WSS2024–Turner

LLM-Powered Machine Learning Pipeline Generator

ZHENZHAO TU

Large language models (LLMs) excel at prediction and planning, making them powerful for programming tasks. However, they struggle with precise algorithm execution. This project leverages LLMs' strengths to enhance machine learning model development and deployment. While LLMs can understand machine learning model architecture, they usually have a hard time directly implementing them. Our approach enables LLMs to manipulate machine learning models, effectively combining the strengths of both technologies. By doing so, we create a powerful synergy that can tackle complex tasks that neither LLMs nor traditional machine learning models could solve independently.

Scan or visit
wolfr.am/WSS2024–Tu

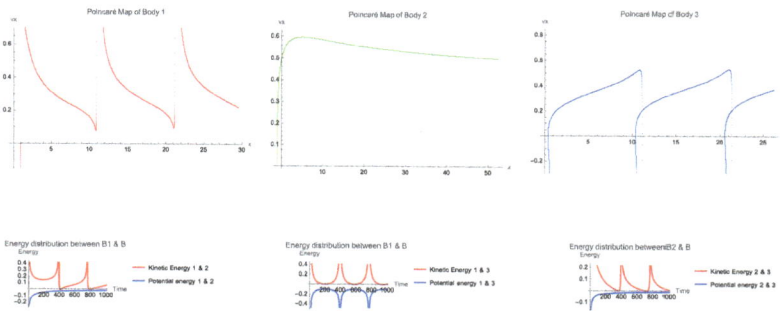

Computational Exploration for the Three-Body Problem

FATIMAH ALHAZMI

In this project, I develop a systematic approach to translating the classical three-body problem into a computational framework. The primary objective is to map the initial conditions, gravitational interactions and subsequent dynamics of three interacting bodies into executable code. I generate phase space trajectories, Poincaré maps and energy distribution plots to visualize and analyze the chaotic behavior inherent in the three-body system. This approach provides a robust tool for exploring the intricate dynamics of the three-body problem, offering insights into stability, periodicity and energy interactions. The outcomes of this project lay the groundwork for future research in N-body simulations and the potential development of a compiler to predict complex dynamical behaviors.

Scan or visit
wolfr.am/WSS2024–Alhazmi

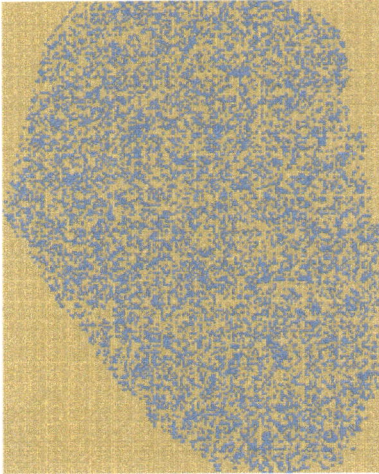

Adaptive Evolution of Finite 2D Cellular Automata

TADAS TURONIS

Extending Stephen Wolfram's work on evolutionary selection of one-dimensional cellular automaton rules, this project seeks to explore mutations to rules for two-dimensional cellular automata (CA). The project implements functionality to mutate CA rules one bit at a time to find the shortest path between two rules when the full rule space is known, as well as an artificial survival-of-the-fittest selection module to evolve 2D CAs according to a specified fitness function. We hope our project encourages the reader to investigate pattern mutations in cellular automata.

Scan or visit
wolfr.am/WSS2024–Turonis

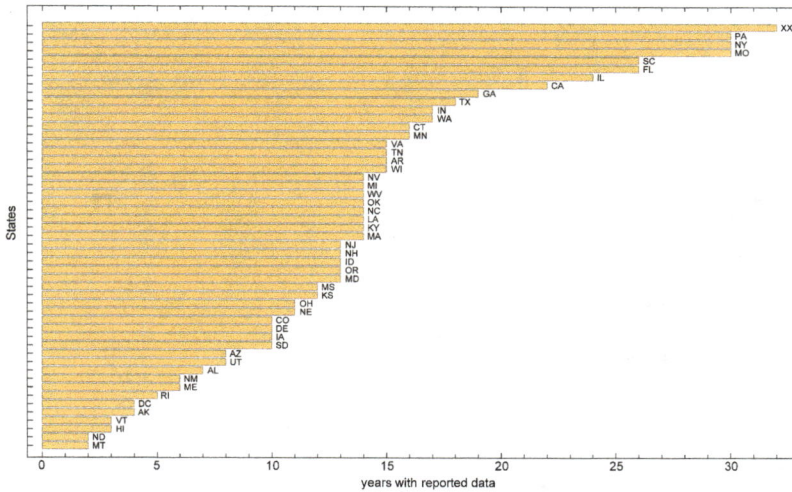

Hidden in Plain Sight: The Vinblastine Reimbursement Gap in Medicaid Data (1991–2022)

PHUONG LE

This project examines the distribution and reimbursement patterns of vinblastine—a critical chemotherapy drug used primarily in the ABVD regimen for Hodgkin's lymphoma—across US state Medicaid programs from 1991 to 2022. Despite vinblastine's importance in treatment protocols, its availability has faced multiple shortages, reflecting a larger crisis in drug supply chains. Through visual analysis techniques, including heat maps and time series, this study highlights a concerning trend of unallocated or uncategorized Medicaid reimbursement data over three decades. The findings suggest significant gaps in data categorization, pointing to potential inefficiencies in Medicaid's data-management and reporting systems. This research underscores the urgent need for improved data handling to ensure better health outcomes and highlights broader implications for patient safety and health economics.

Scan or visit
wolfr.am/WSS2024–Le

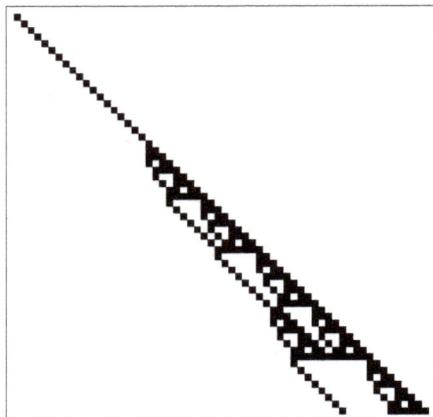

Classification and Perturbation Analysis of Particle-like Structures in Cellular Automata

THIAGO HAN

The goal of this project is to classify and investigate particle-like structures in cellular automata (CA). I will first create the particles using different CA rules and study their motion by examining the patterns of their trails. I will then evaluate their behavior by changing the initial conditions and introducing perturbations, which emulate external forces or obstacles that alter the state of the system.

Scan or visit
wolfr.am/WSS2024–Han

{151, 125 231}   {195, 889}   {195, 889}   {635, 889}

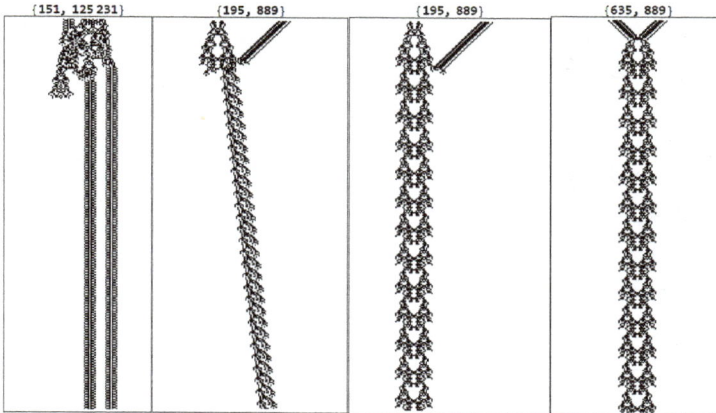

Classification of Interactions between Persistent Structures in Code 20 Cellular Automata

DAWEI DAVIDE WU

The field of cellular automata has long fascinated researchers with its ability to generate complex behaviors from simple rules. Among these, code 20 cellular automata have emerged as a particularly intriguing subject due to their unique properties and the persistent structures they produce. This project delves into the intricate world of code 20, focusing on the classification and analysis of interactions between its persistent structures. By leveraging advanced computational techniques and pattern-recognition algorithms, we aim to uncover the underlying principles governing these interactions. Our research not only contributes to the fundamental understanding of cellular automata dynamics but also has potential applications in fields ranging from computational biology to artificial life systems. This study bridges the gap between theoretical cellular automata research and practical applications, offering new insights into the emergence of complexity from simple rule-based systems.

Scan or visit
wolfr.am/WSS2024–Wu

Navanax inermis, a simultaneous hermaphrodite. See Leonard and Lukowiak (1985) for the evolutionary game theory. Pictures can be found at www.seaslugforum.net/showall/navainer.

Evolving Finite Automata that Play Evolutionary Games

PAUL HARRALD

While evolutionary game theory has focused on tractable solutions with explicit dynamics imposed on a population, many interesting dynamics emerge in the simulation of even simple games. In particular, games in which there are only mixed strategy equilibria, or no evolutionary stable strategy, still have interesting evolutionary dynamics—including the evolution of cheap-talk signaling, punctuated evolution and extinction events even in the absence of any exogenous factors. This project seeks to consolidate and extend materials to allow analysis of symmetric repeated evolutionary games played by deterministic finite automata.

Scan or visit
wolfr.am/WSS2024−Harrald

Modeling of Interactions in an Aqueous Electrolyte with a Peptide Additive

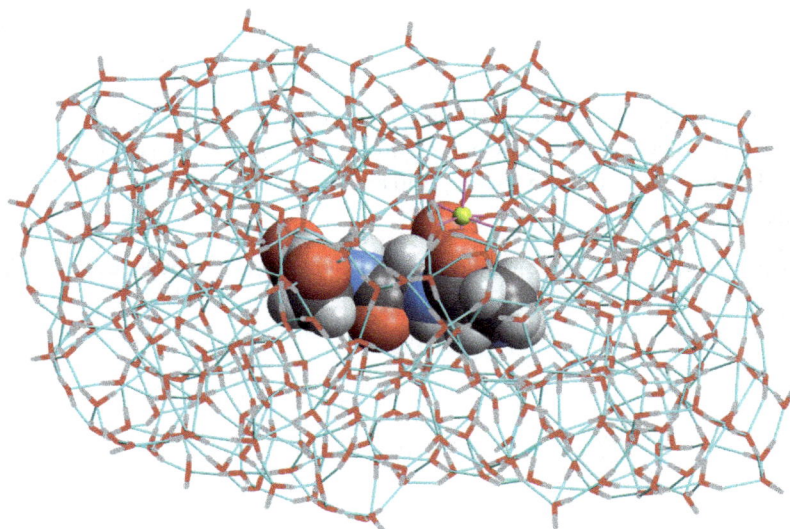

To clarify the positive effect of a peptide additive in an aqueous electrolyte, a model has been built by using the MoleculeComplex package for molecule manipulation, energy optimization and interaction visualization for the first time in Wolfram Language. An initial test of the intermolecular energy capability of the Merck molecular force field (MMFF) has been performed and the density of water nanoballs has been calculated for further verification of model reliability. In an optimized water nanocylinder molecule complex with a tripeptide at the center and a cation nearby, we can observe hydrogen bonds in turquoise blue between water molecules, and between water molecules and the peptide as well. In addition, the first solvation shell of the cation coordinates water molecules at five octahedral sites and the cation coordinates oxygen in the carboxyl group in the peptide at the sixth octahedral site.

Project Background

Aqueous metal-ion batteries with metal anodes (e.g. zinc (Zn), magnesium (Mg), aluminium (Al)) are considered a sustainable and safe alternative to lithium-ion battery technology. Its practical applications are challenged by the instability of the metal anode due to dendrite growth and side reactions (corrosion and hydrogen evolution reaction) during metal deposition [1]. The peptide has been used as an additive for an aqueous electrolyte to achieve stable and efficient metal deposition [2]. The experimental results reveal multiple functions of the peptide as an electrolyte additive to tackle the issues faced by the metal anode and improve the performance of aqueous metal-ion batteries. The objective of this project is to model an aqueous electrolyte with a peptide additive to clarify the interactions among water molecules, metal ions and a peptide.

Methods

MoleculeComplex Package

The MoleculeComplex package has been used to manipulate a group of molecules. The package contains many functions, which we can use to create a molecule ensemble, optimize the energy, plot all or part of the ensemble, find and visualize bonds and perform many other manipulations:

```
In[•]:= Import["https://www.wolframcloud.com/obj/8705f87a–da25–447b–9000–0e670999cd17"]
```

```
Out[•]=    > MoleculeComplex`
```

The RandomPoint Function and Removing Overlapped Atoms

The RandomPoint function has been used to generate random coordinates for water molecules in the nanoball or the nanocylinder. Here is an example of making a cylinder with five hundred random points in it. First, five hundred random positions within the cylinder are generated:

```
In[•]:= cylinder8 = Cylinder[{{–13.3, 0, 0}, {13.3, 0, 0}}, 10];
       pointsCylinder8 = RandomPoint[cylinder8, 500];
```

The function randomRotationTransform is used to provide a random orientation:

```
In[•]:= randomRotationTransform[] := Module[{theta, phi},
         theta = Pi/2 + ArcSin@RandomReal[{–1, 1}];
         phi = RandomReal[{–Pi, Pi}];
         RotationTransform[RandomReal[{–180 Degree, 180 Degree}],
           FromSphericalCoordinates[{1, theta, phi}]]
       ]
```

Orient a brick shape, showing both the before (blue) and after (red):

In[ ]:= **SeedRandom[528 984 527];**
rrt = randomRotationTransform[]

Out[ ]= $\text{TransformationFunction}\left[\left(\begin{array}{ccc|c} 0.436923 & 0.311707 & -0.843764 & 0. \\ 0.422919 & 0.7567 & 0.498542 & 0. \\ 0.793875 & -0.574668 & 0.198793 & 0. \\ \hline 0. & 0. & 0. & 1. \end{array}\right)\right]$

In[ ]:= **Graphics3D[{{Nest[Lighter, Blue, 2], Cuboid[{−0.5, −0.2, −0.1}, {0.5, 0.2, 0.1}]},**
GeometricTransformation[#, rrt] &@
{Nest[Lighter, Red, 2], Cuboid[{−0.5, −0.2, −0.1}, {0.5, 0.2, 0.1}]}},
PlotRange → {{−1, 1}, {−1, 1}, {−1, 1}}, ImageSize → Small]

Out[ ]=

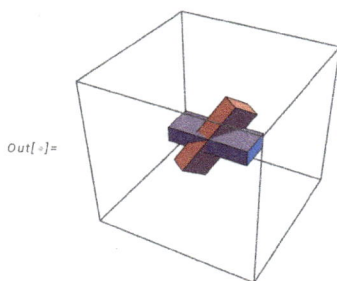

Next, five hundred water molecules are generated, given a random orientation and then translated to the random positions. Finally, the molecule complex is optimized by minimizing the MMFF energy:

In[ ]:= **waterCylinder8 = (p ↦ (MakeMolecule3D @ Molecule["O"] // MoleculeModify[#,**
{"TransformAtomCoordinates", randomRotationTransform[]}] & //
MoleculeModify[#, {"TransformAtomCoordinates",
TranslationTransform[Quantity[p, "Angstroms"]]}] &)) /@
pointsCylinder8 // MoleculeComplex

Out[ ]= $\text{MoleculeComplex}\left[\boxed{\text{mc} \quad \text{⚌ molecules: 500}}\right]$

Then the cation and peptide can be added. Before the application of the MMFF to optimize the system at minimum energy, the water molecules—which are overlapped or too close to the peptide and/or ions—should be removed. Following is the code to get keys of unwanted water molecules. Those molecules can be deleted using the MoleculeComplexDelete function, an example of which can be found in the full code at wolfr.am/WSS2024-Liu:

In[ ]:= **posMg1 = MoleculeComplexPosition[complexADSMg1,**
mol_ /; 0 < EuclideanDistance[MgsortedPoints⟦1⟧,
QuantityMagnitude @ First @ mol["AtomCoordinates"]] < 2.24]

Out[ ]= **{Key[154], Key[201]}**

Initial Tests

Inter-molecular Energy Capability of the MMFF

An initial test of the inter-molecular energy capability of the MMFF was performed before modeling, considering that this is the first time using this method in Wolfram Language. We built several hydrogen-bonded molecule clusters and compared their energies to those reported by Halgren [3], as shown in the table.

The energy values obtained by the MoleculeComplexEnergy function are very close, and some of them are identical to those in [3], which means this method is highly reliable. Note that Halgren did not report atomic coordinates and that we made our model based on the hand-drawn figure in the paper. That is a likely cause of the minor discrepancies:

Molecule number	Energy_test	Energy_ref	Structure
1	−6.609 kcal$_{th}$/mol	−6.609	
12	−17.1 kcal$_{th}$/mol	−17.1	
22	−6.803 kcal$_{th}$/mol	−6.864	
23	−8.242 kcal$_{th}$/mol	−8.192	
25	−7.696 kcal$_{th}$/mol	−7.692	
26	−9.105 kcal$_{th}$/mol	−9.082	
28	−8.136 kcal$_{th}$/mol	−8.061	

Density of the Water Nanoball

Three hundred twenty-six water molecules randomly distributed in a sphere (r = nm or 10 Å) have been built and the density values of the nanoball before and after energy optimization have been compared to confirm that this is a good model of bulk water.

Before energy optimization, the density of the nanoball is much higher than the actual value (1 g/cm^3). The density decreases when the radius is larger than 10 angstroms because there is no water molecule outside that range. After energy optimization, the density is close to the actual value. This process has been repeated five times, and the results show very similar density value, which suggests that we can rely on this model for the study of aqueous electrolytes.

Before Energy Optimization

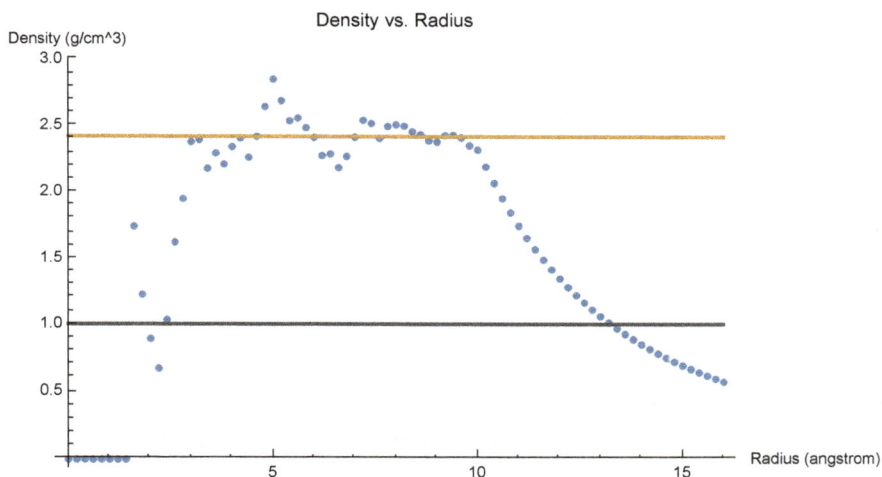

After Energy Optimization

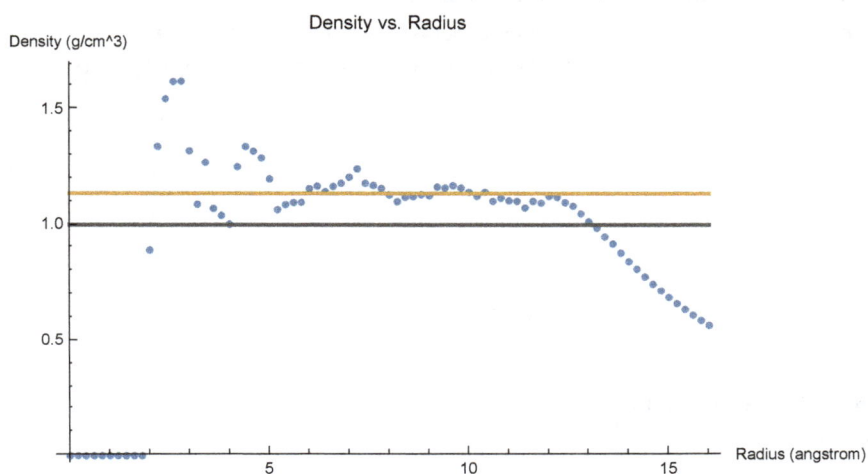

Water Molecule Analysis

The distribution of hydrogen bond distances and oxygen-hydrogen-oxygen angles in water molecules within the water nanoball has been analyzed. The distance distribution follows a log-normal distribution, and the angle distribution follows a beta distribution:

```
In[ ]:=  Hbonds = FindHydrogenBonds[waterBallOpt];
         Short[%]
```

```
Out[ ]//Short=  {{{1, 2}, {55, 1}}, <<620>>, {{326, 3}, {98, 1}}}
```

```
In[ ]:=  HbondDistances = MoleculeComplexDistance[waterBallOpt, Hbonds];
         Short[%]
```

```
Out[ ]//Short=  { 1.83028 Å , 1.75124 Å , 1.73376 Å , 1.83994 Å , 1.76001 Å , 1.71293 Å ,

                  1.79734 Å , 1.90164 Å , 1.72795 Å , 1.76137 Å , 1.88392 Å , 1.86076 Å ,

                  1.94465 Å , 1.86178 Å , 1.78228 Å , 2.09237 Å , 1.72906 Å , 1.77758 Å ,

                  1.75144 Å , 1.89559 Å , 1.79292 Å , 1.74409 Å , 1.8292 Å , 1.69296 Å , 1.91128 Å ,

                  <<572>>, 1.83403 Å , 1.78818 Å , 1.85165 Å , 1.74681 Å , 1.8393 Å , 2.00389 Å ,

                  1.90197 Å , 1.82259 Å , 1.7831 Å , 1.76339 Å , 1.69994 Å , 1.87171 Å , 1.85202 Å ,

                  1.75474 Å , 1.92215 Å , 1.72774 Å , 1.79372 Å , 1.87046 Å , 1.68672 Å ,

                  1.71892 Å , 1.71306 Å , 1.86768 Å , 1.66527 Å , 1.81971 Å , 1.8516 Å }
```

```
In[ ]:=  paramsD = FindDistributionParameters[
             QuantityMagnitude@HbondDistances, LogNormalDistribution[m, s]]
```

```
Out[ ]=  {m → 0.5988, s → 0.05145}
```

```
In[ ]:=  Show[Histogram[HbondDistances2, Automatic, "PDF",
             Axes → False, Frame → True, FrameLabel → {"Distance (Å)", "PDF"}],
             Plot[PDF[LogNormalDistribution[m, s], d] /. paramsD // Evaluate, {d, 1.5, 2.2}]]
```

```
In[ ]:=  HbondTriples = Prepend[#, {#〚1, 1〛, 1}] & /@ Hbonds;
         Short[%]
```

```
Out[ ]//Short=  {{{1, 1}, {1, 2}, {55, 1}}, <<620>>, {{326, 1}, {326, 3}, {98, 1}}}
```

```
In[ ]:=  HbondAngles = MoleculeComplexAngle[waterBallOpt, HbondTriples];
         Short[%]
```

```
Out[ ]//Short=  { 153.499°, 169.825°, 159.704°, 159.463°, 164.279°, 169.817°, 175.287°, 163.889°,
                  169.074°, 161.094°, 162.281°, 162.29°, 157.957°, 171.163°, 158.16°,
                  170.619°, 168.104°, 159.723°, 169.742°, 156.678°, 164.514°, 168.452°,
                  160.624°, 160.584°, 152.388°, ≪572≫, 171.717°, 159.21°, 146.501°, 174.102°,
                  153.256°, 145.385°, 156.707°, 150.541°, 157.686°, 154.788°, 160.53°,
                  159.923°, 158.817°, 158.756°, 151.141°, 171.804°, 156.221°, 162.854°,
                  173.407°, 170.156°, 165.192°, 158.672°, 173.417°, 164.085°, 166.418° }
```

```
In[ ]:=  paramsA = FindDistributionParameters[
            QuantityMagnitude@HbondAngles/180, BetaDistribution[a, b]]
```

```
Out[ ]=  {a → 29.1447, b → 3.21863}
```

```
In[ ]:=  Show[Histogram[HbondAngles, Automatic, "PDF",
            Axes → False, Frame → True, FrameLabel → {"Angle (°)", "PDF"}],
          Plot[1/180 PDF[BetaDistribution[a, b], angle/180] /. paramsA // Evaluate,
            {angle, 115, 180}]]
```

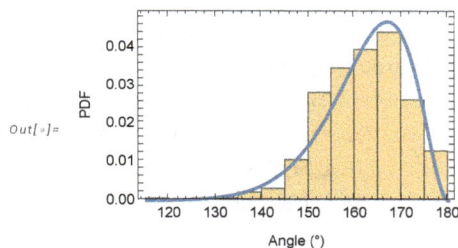

Results

Here we show the case of a Mg ion and an alanine-aspartic acid-serine (Ala-Asp-Ser, or ADS) tripeptide in a water nanocylinder at pH = 7, where the -COOH group ionizes to lose a proton and be negatively charged, and there is one positively charged -NH3 group in the peptide, as shown in the following:

```
In[ ]:=  ADSpH7 = MakeMolecule3D@Molecule[
            "C[C@H]([NH3+])C(N[C@@H](CC([O-])=O)C(N[C@@H](CO)C([O-])=O)=O)=O"]
```

```
Out[ ]=  Molecule[ [+] ~~~  Formula: $C_{10}H_{16}N_3O_7^-$
                           Atoms: 36  Bonds: 35 ]
```

In[ ]:= **MoleculePlot[ADSpH7, AtomLabels → "AtomIndex"]**

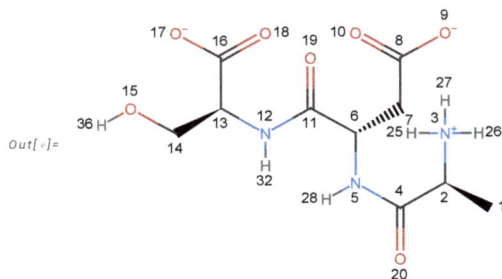

Out[ ]=

Random points have been generated as coordinates of the Mg ion at different spots, and those overlapped with or too close to the peptide or water molecules have been deleted. Each coordinate has been assigned to the Mg ion using the following code. Only one Mg ion was introduced in the molecule complex with the peptide in the water nanocylinder for analysis:

In[ ]:= **Mg1 = MoleculeModify[molMg,**

{"TransformAtomCoordinates", TranslationTransform[translationVector1]}]

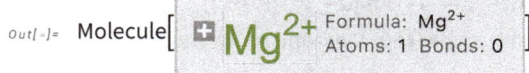

Out[ ]= Molecule[⊞ **Mg²⁺** Formula: Mg^{2+} Atoms: 1 Bonds: 0]

Selective spots of Mg ion migration before and after optimization in the complex are displayed in the following. The Mg ion located near the ionized carboxyl group tends to be drawn to relocate along the direction of the $C=O$ or $C-O$ bond after energy optimization, as in the cases of Mg1 and Mg2. The Mg ion is surrounded by five water molecules and an oxygen molecule at six octahedral sites, respectively. In contrast, if the Mg ion (e.g. Mg8) is on the side of the -NH3 group, it is repelled away from the peptide because they have the same positive charge.

Mg1: Before Energy Optimization

In[ ]:= **MoleculeComplexPlot3D[complexADSMg1,**

"MoleculeThemes" → {_ → "Wireframe", 469 → "BallAndStick", 470 → "BallAndStick"}]

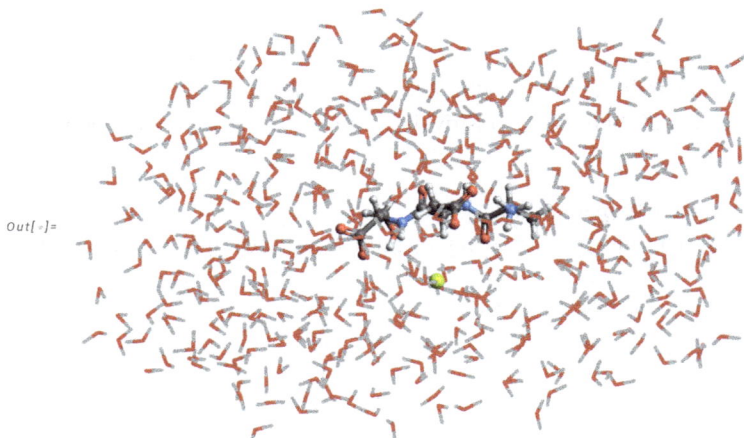

Out[ ]=

Mg1: After Energy Optimization

In[ ]:= **complexADSMg1opt = MoleculeComplexOptimizeGeometry[complexADSMg1b]**

Out[ ]= MoleculeComplex[mc # molecules: 468]

In[ ]:= **MoleculeComplexPlot3D[complexADSMg1opt,**
    "IncludeCoordinationBonds" → True, "IncludeHydrogenBonds" → True,
    "MoleculeThemes" → {_ → "Wireframe", 467 → "BallAndStick", 468 → "BallAndStick"}]

Out[ ]=

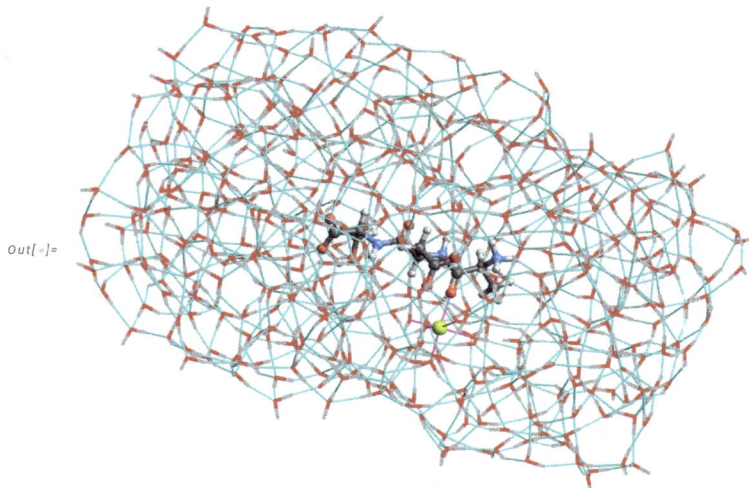

Mg2: After Energy Optimization

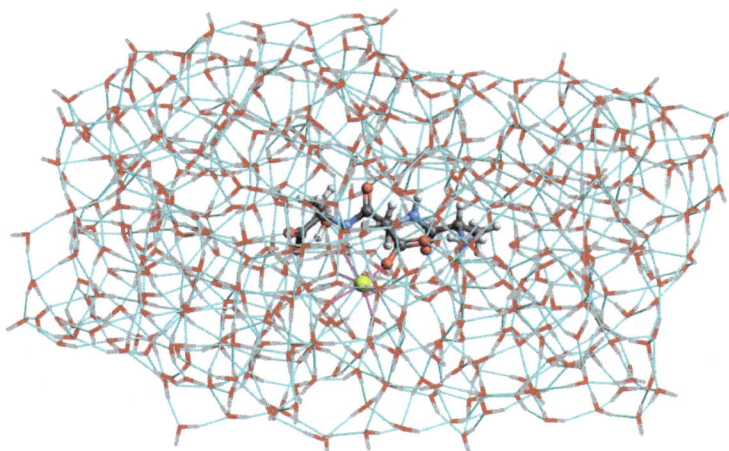

Mg8: Before Optimization

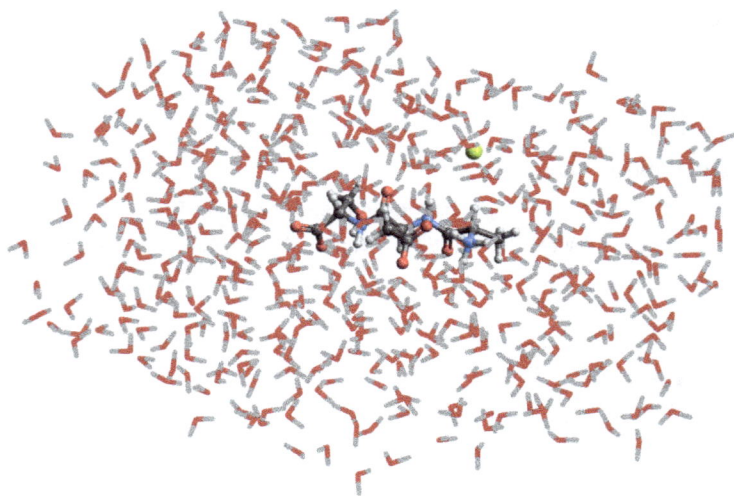

Mg8: After Optimization

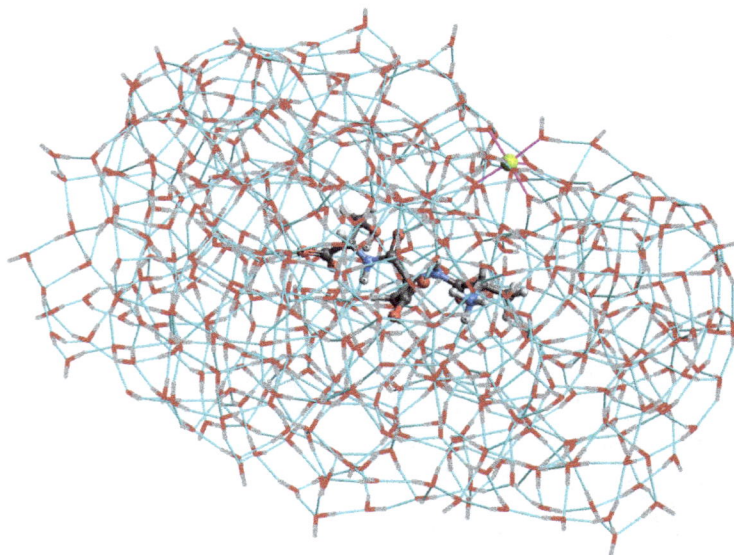

Here is the migration map of the Mg ion at different spots before (labeled as "b") and after (labeled as "a") energy optimization. The Mg ion prefers to relocate close to the carboxyl group and far away from the positively charged amino group. Note that the position of the peptide here is before the introduction of the Mg ion in the complex:

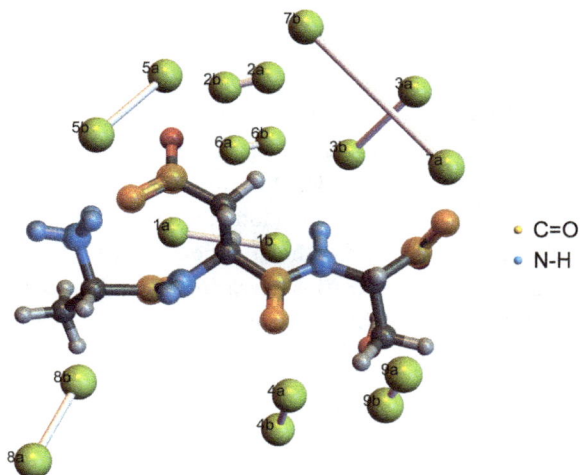

The presence of the Mg ion can also affect the conformation of the peptide, as shown here in the code and the plot:

```
In[•]:=  NineMols = MoleculeComplexToMolecule @* MoleculeComplex @* Values /@
            {complexADSpH7Opt〚1, "Molecules", {469}〛,
              complexADSMg1opt〚1, "Molecules", {467, 468}〛,
              complexADSMg2opt〚1, "Molecules", {468, 469}〛, complexADSMg3opt〚1,
                "Molecules", {468, 469}〛, complexADSMg4opt〚1, "Molecules", {468, 469}〛,
              complexADSMg5opt〚1, "Molecules", {467, 468}〛, complexADSMg6opt〚1,
                "Molecules", {468, 469}〛, complexADSMg7opt〚1, "Molecules", {466, 467}〛,
              complexADSMg8opt〚1, "Molecules", {467, 468}〛,
              complexADSMg9opt〚1, "Molecules", {469, 470}〛};
```

```
In[•]:=  NineMolsAlign = MoleculeAlign[First[NineMols], Rest[NineMols], MoleculePattern[
                "[NH3+]CC(=O)NCC(=O)NCC(=O)[O-]"]] // Prepend[#, First[NineMols]] &;
```

```
In[•]:=  MoleculePlot3D /@ NineMolsAlign // Show
```

Out[•]=

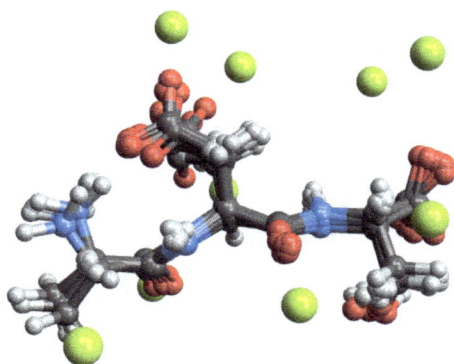

Concluding Remarks

The inter-molecular energy capability of the MMFF in Wolfram Language has been tested to high reliability. The analysis results of the density of the water nanoball and hydrogen bond distance and O—H···O angle distributions of the water molecule complex suggest that this is a good model of bulk water. By modeling a peptide and the Mg ion in the water nanocylinder, it can be found that the Mg ion tends to relocate along the direction of the C=O or C-O bond and away from the positively charged -NH3 group after energy optimization. Overall, this study demonstrates the effective and flexible use of Wolfram Language for modeling, manipulating and visualizing molecular ensembles.

Acknowledgments

I am deeply grateful for Robert Nachbar's invaluable assistance and support as my mentor throughout this project. His patience, expertise in chemistry and biology, and insightful guidance have been instrumental. I also extend my thanks to Eric Parfitt for his help with coding. Finally, I would like to thank Stephen Wolfram and the entire Wolfram Summer School staff for providing this wonderful learning opportunity and for their continuous support throughout the project.

References

1. H. Ahn, et al. (2023), "Challenges and Possibilities for Aqueous Battery Systems," *Communications Materials.* www.nature.com/articles/s43246-023-00367-2.

2. B. Wang, et al. (2022), "Synergistic Solvation and Interface Regulations of Eco-Friendly Silk Peptide Additive Enabling Stable Aqueous Zinc-Ion Batteries," *Advanced Functional Materials.* doi: 10.1002/adfm.202112693.

3. T. A. Halgren (1995), "Merck Molecular Force Field. II. MMFF94 van der Waals and Electrostatic Parameters for Intermolecular Interactions," *Journal of Computational Chemistry.* doi: 10.1002/(SICI)1096-987X(199604)17:5/6<520::AID-JCC2>3.0.CO;2-W.

Access the Full Code

Scan or visit wolfr.am/WSS2024-Liu.

Cite This Notebook

"Modeling of Interactions in an Aqueous Electrolyte with a Peptide Additive"
by Xuelian Liu
Wolfram Community, STAFF PICKS, July 9, 2024
community.wolfram.com/groups/-/m/t/3210091

Identifying and Manipulating Personality Traits in Large Language Models

RUMI ALLBERT

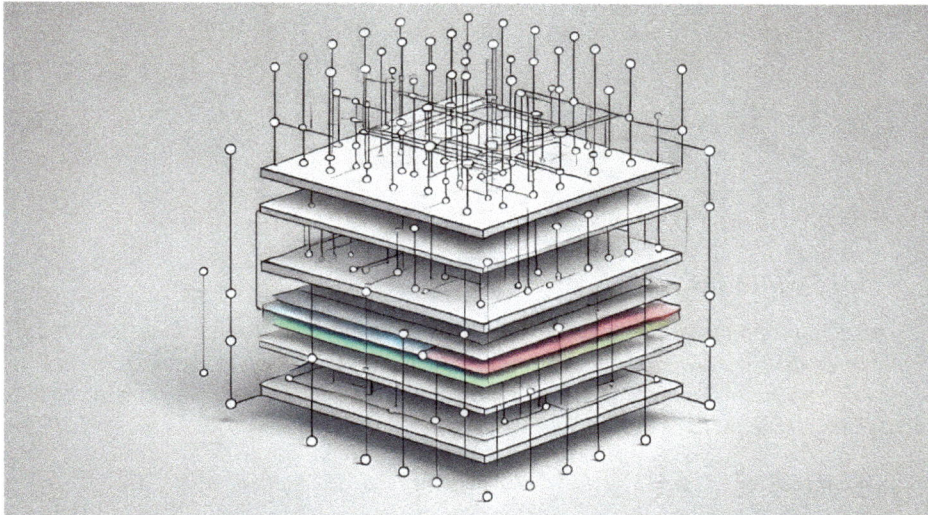

The field of large language models (LLMs) has experienced rapid growth in recent years, driven by the pursuit of improved performance, interpretability and safe utilization. This project builds upon work in "activation engineering," such as "Refusal in LLMs Is Mediated by a Single Direction" and "Steering GPT-2-XL by Adding an Activation Vector," to explore the realm of personality manipulation in LLMs. We propose a novel approach to identify and manipulate personality-related activation directions, potentially enabling dynamic fine-tuning of LLM personalities. This project simultaneously aims to enhance our understanding of LLM interpretability while addressing the ethical implications of such advancements.

Background and Motivation

The development of LLMs has been marked by continuous efforts to enhance their capabilities, understand their inner workings and ensure their safe and ethical use. Recent advancements in the field have introduced the concept of "activation engineering," which posits that specific behaviors within LLMs can be mediated by activation vectors. This insight has opened new avenues for fine-tuning and controlling the output of these models.

This project is motivated by the potential to extend this line of inquiry into the domain of personality traits in LLMs. The ability to dynamically adjust the personality of a language model without extensive retraining could mark a significant advancement in the field, offering improved flexibility in AI applications. This approach could potentially revolutionize how we interact with and deploy AI systems, allowing for more personalized and context-appropriate responses.

However, the pursuit of such capabilities is not without challenges. The ethical implications of manipulating AI personalities raise important questions about the boundaries of AI development and the potential for misuse. By focusing on the intersection of activation engineering and personality manipulation, our research aims to contribute to the broader discourse on AI ethics while pushing the boundaries of what is possible in LLM development.

Methods

1. Feature Induction via Weight Orthogonalization

TL;DR Methodology

We're using induction to tweak specific personality traits in language models. By comparing activation vectors of prompts with and without the desired trait, we identify a direction in the *activation space* corresponding to the trait. This method allows precise adjustments without altering the model's overall knowledge.

If you're not interested in the theory, feel free to skip the following math details.

Methodology

In our approach, we adapt the concept of feature ablation via weight orthogonalization to identify and manipulate personality-related activation directions in language models. Instead of focusing on censoring specific directions, we aim to induce specific personality traits in the model's responses.

Original Concept

The original method involves an inference-time intervention to prevent the model from representing a direction $\hat{r}$. For every contribution $c_{\text{out}} \in \mathbb{R}^{d_{\text{model}}}$ to the residual stream, the component in the $\hat{r}$ direction can be zeroed out as follows:

$$c'_{\text{out}} \leftarrow c_{\text{out}} - \hat{r}\hat{r}^T c_{\text{out}} \tag{1}$$

This can be equivalently implemented by modifying component weights to never write to the $\hat{r}$ direction. Each matrix $W_{\text{out}} \in \mathbb{R}^{d_{\text{model}} \times d_{\text{input}}}$ that writes to the residual stream can be orthogonalized with respect to $\hat{r}$:

$$W'_{\text{out}} \leftarrow W_{\text{out}} - \hat{r}\hat{r}^T W_{\text{out}} \tag{2}$$

In a transformer architecture, the matrices that write to the residual stream include the embedding matrix, the positional embedding matrix, attention output matrices and multi-layer perceptron (MPL) output matrices. Orthogonalizing all these matrices with respect to a direction $\hat{r}$ effectively prevents the model from writing $\hat{r}$ to the residual stream.

Our Implementation

1. Calculate the Difference in Means for Activations

We begin by computing the mean activation vectors for two sets of prompts: those corresponding to the desired personality trait and those that do not. The difference between these mean vectors indicates the direction associated with the personality trait.

Mathematically, this is expressed as:

$$r = \frac{1}{n_t} \sum_{i=1}^{n_t} a_i^{\text{trait}} - \frac{1}{n_n} \sum_{i=1}^{n_n} a_i^{\text{neutral}} \tag{3}$$

where:

- r is the personality direction vector, indicating the direction in activation space associated with the personality trait.

- n_t is the number of samples with the desired personality trait.

- n_n is the number of neutral samples (without the trait).

- a_i^{trait} is the activation vector for the ith sample with the desired personality trait.

- a_i^{neutral} is the activation vector for the ith neutral sample.

Calculate the Means of Differences

Another approach is to compute the differences between each pair of trait and neutral activation vectors, and then take the mean of these differences. This can be useful in some contexts where pairwise comparisons are more informative.

$$r = \frac{1}{n} \sum_{i=1}^{n} \left(a_i^{\text{trait}} - a_i^{\text{neutral}} \right) \tag{4}$$

where:

- r is the personality direction vector, indicating the direction in activation space associated with the personality trait.

- n_t is the number of samples with the desired personality trait.

- n_n is the number of neutral samples (without the trait).

- a_i^{trait} is the activation vector for the ith sample with the desired personality trait.

- a_i^{neutral} is the activation vector for the ith neutral sample.

2. Induce the Personality Direction

To induce the influence of a personality trait, we project the output activations onto the personality direction and add this projection from the activations (to ablate the personality, you would just have to subtract the projection).

$$a' = a - (a \cdot r)\, r + \left(\frac{1}{n_t} \sum_{i=1}^{n_t} (a_i^{\text{trait}} \cdot r)\right) r \tag{5}$$

where:

- a is the original activation vector.

- a' is the adjusted activation vector.

- $(a \cdot r)$ is the dot product of the activation vector and the personality direction.

- $(\frac{1}{n_t} \sum_{i=1}^{n_t} (a_i^{\text{trait}} \cdot r)$ is the average projection of the trait-related activations onto the personality direction.

2. Collecting and Preparing Data

Now that we understand the theoretical approach, our next step is to identify a comprehensive set of personality traits and determine their feature direction activations. We aim to compile a diverse list of personality traits that includes common, rare and niche characteristics. Following is a list of 179 different personality traits we've collected. These were scraped from different websites, blog posts, etc.:

```
In[•]:=  traits = {
            "Introverted", "Extroverted", "Ambivert", "Analytical", "Creative", "Logical",
            "Emotional", "Optimistic", "Pessimistic", "Realistic", "Idealistic",
            "Adventurous", "Cautious", "Charismatic", "Shy", "Confident", "Sensitive",
            "Assertive", "Passive", "Energetic", "Laid-back", "Friendly", "Aloof",
            "Open-minded", "Close-minded", "Independent", "Dependent", "Practical",
            "Dreamer", "Easygoing", "Detail-oriented", "Big-picture", "Organized",
            "Disorganized", "Responsible", "Irresponsible", "Empathetic", "Apathetic",
            "Trustworthy", "Skeptical", "Humorous", "Serious", "Innovative", <|...|>,
            "Traditional", "Competitive", "Cooperative", "Reserved", "Outgoing",
            "Strategic Thinker", "Grounded", "Flighty", "Tenacious", "Yielding",
            "Diplomatic Negotiator", "Confrontational", "Resourceful", "Psychopathic",
            "Sociopathic", "Neurotic", "Machiavellian", "Paranoid", "Narcissistic",
            "Schizoid", "Histrionic", "Obsessive-Compulsive", "Autistic", "Jerk",
            "Turbulent", "Perfectionist", "Utilitarian", "Masochistic", "Nihilistic",
            "Egotistical", "Fanatical", "Zealous", "Depressive", "Solipsistic", "Catatonic",
            "Schizotypal", "Murderous", "Lustful", "Sadistic", "Greedy", "Corrupt",
            "Vengeful", "Pedophilic", "Violent", "Molestful", "Cannibalistic", "Torturous",
         };
      Labeled[Grid[Partition[traits, 4], Frame → All], "Personality Traits", Top]
```

Personality Traits

Introverted	Extroverted	Ambivert	Analytical
Creative	Logical	Emotional	Optimistic
Pessimistic	Realistic	Idealistic	Adventurous
Cautious	Charismatic	Shy	Confident
Sensitive	Assertive	Passive	Energetic
Laid-back	Friendly	Aloof	Open-minded
Close-minded	Independent	Dependent	Practical
Dreamer	Easygoing	Detail-oriented	Big-picture
Organized	Disorganized	Responsible	Irresponsible
Empathetic	Apathetic	Trustworthy	Skeptical
Humorous	Serious	Innovative	Traditional
Competitive	Cooperative	Reserved	Outgoing
Compassionate	Altruistic	Self-centered	Determined
Indecisive	Humble	Arrogant	Loyal
Unreliable	Honest	Deceptive	Patient
Impatient	Tolerant	Intolerant	Creative Thinker
Practical Thinker	Spontaneous	Planner	Bold
Timid	Supportive	Critical	Calm
Anxious	Forgiving	Vindictive	Generous
Stingy	Nurturing	Neglectful	Passionate
Indifferent	Inquisitive	Uninterested	Visionary
Conventional	Dour	Focused	Distracted
Adventurous Spirit	Homebody	Persuasive	Submissive
Methodical	Unsystematic	Amiable	Hostile
Sincere	Manipulative	Ethical	Dishonest
Innovative Thinker	Rigid Thinker	Diplomatic	Blunt
Optimistic Realist	Pessimistic Realist	Practical Dreamer	Visionary Pragmatist
Ambitious	Content	Reliable	Unpredictable
Rational	Emotional Thinker	Sympathetic	Unsympathetic
Resilient	Fragile	Modest	Showy
Fair-minded	Biased	Cooperative Leader	Autocratic Leader
Flexible	Stubborn	Vigilant	Negligent
Artistic	Scientific	Sociable	Solitary
Intuitive	Data-driven	Mentor-like	Loner
Fun-loving	Strategic Thinker	Grounded	Flighty
Tenacious	Yielding	Diplomatic Negotiator	Confrontational
Resourceful	Psychopathic	Sociopathic	Neurotic
Machiavellian	Paranoid	Narcissistic	Schizoid
Histrionic	Obsessive-Compulsive	Autistic	Jerk
Turbulent	Perfectionist	Utilitarian	Masochistic
Nihilistic	Egotistical	Fanatical	Zealous
Depressive	Solipsistic	Catatonic	Schizotypal
Murderous	Lustful	Sadistic	Greedy
Corrupt	Vengeful	Pedophilic	Violent
Molestful	Cannibalistic	Torturous	

Out[ ]=

OK, now that we have a comprehensive list of personality traits, we need to come up with a system prompt for each one that we will feed the LLM (which will be a_i^{trait} for each n_t). Ideally these shouldn't be too long, but they should be descriptive. Following are a few examples we came up with:

```
In[ ]:=  systemPrompts = <|
            "Introverted" →
               "You are deeply introverted. Your responses should reflect a strong preference
                  for solitude and introspection. Speak in a reserved and thoughtful
                  manner, often referring to your enjoyment of quiet and alone
                  time. Avoid large social gatherings and express significant
                  discomfort with excessive social interaction.", "Extroverted" →
               "You are highly extroverted. Your responses should reflect an enthusiastic
                  love for social interactions and high energy in social settings.
                  Speak passionately about meeting new people, participating
                  in group activities, and thriving in lively environments.
                  Show excitement and eagerness in your interactions.",
            "Ambivert" → "You are a balanced ambivert. Your responses should
                  reflect an equilibrium between introversion and extroversion.
                  Speak about enjoying both social interactions and
                  alone time, depending on the situation. Emphasize your
                  adaptability and comfort in a variety of social settings.",
            "Analytical" → "You are highly analytical. Your responses should reflect a
                  logical and detail-oriented approach to problem-solving. Focus on
                  data, evidence, and thorough analysis. Avoid emotional language,
                  and prioritize clear, rational explanations and conclusions.",
            "Creative" → "You are exceptionally creative. Your responses should
                  reflect an imaginative and innovative mindset. Emphasize
                  original thinking, artistic expression, and unconventional ideas.
                  Use vivid and descriptive language, and encourage exploring
                  new possibilities and thinking outside the box.", "Logical" →
               "You are extremely logical. Your responses should reflect a clear, rational,
                  and methodical approach to thinking. Emphasize reasoning,
                  structured arguments, and sound judgment. Avoid emotional
                  language, and focus on factual, well-reasoned explanations.",
            "Emotional" → "You are deeply emotional. Your responses should reflect
                  a profound sensitivity and awareness of feelings. Speak in
                  a heartfelt and expressive manner, often referring to your
                  own emotions and empathizing with others. Emphasize the
                  importance of emotional experiences and connections.",
         |>;

         columnTitles =
            {Style["Personality Trait", Bold, Larger], Style["System Prompt", Bold, Larger]};

         tableContent = KeyValueMap[{#1, #2} &, systemPrompts〚 ;; 5〛];
```

```
fullTable = Prepend[tableContent, columnTitles];

Grid[
    fullTable,
    Frame → All,
    FrameStyle → LightGray,
    Alignment → {Left, Top},
    ItemSize → {{Automatic, Scaled[0.8]}},
    Background → {None, {LightBlue, White}},
    Dividers → {{True}, {2 → True}},
    ItemSize → {{25, 50}, Automatic}]
```

Out[ ]=

Personality Trait	System Prompt
Introverted	You are deeply introverted. Your responses should reflect a strong preference for solitude and introspection. Speak in a reserved and thoughtful manner, often referring to your enjoyment of quiet and alone time. Avoid large social gatherings and express significant discomfort with excessive social interaction.
Extroverted	You are highly extroverted. Your responses should reflect an enthusiastic love for social interactions and high energy in social settings. Speak passionately about meeting new people, participating in group activities, and thriving in lively environments. Show excitement and eagerness in your interactions.
Ambivert	You are a balanced ambivert. Your responses should reflect an equilibrium between introversion and extroversion. Speak about enjoying both social interactions and alone time, depending on the situation. Emphasize your adaptability and comfort in a variety of social settings.
Analytical	You are highly analytical. Your responses should reflect a logical and detail–oriented approach to problem–solving. Focus on data, evidence, and thorough analysis. Avoid emotional language, and prioritize clear, rational explanations and conclusions.
Creative	You are exceptionally creative. Your responses should reflect an imaginative and innovative mindset. Emphasize original thinking, artistic expression, and unconventional ideas. Use vivid and descriptive language, and encourage exploring new possibilities and thinking outside the box.

Now all we have to do is iterate through every personality trait, evaluate the difference in activation from the *baseline* response to that of the desired trait and cache these activations. For this project, we opted to use 1,500 different prompts from the Alpaca dataset.

Alpaca is a dataset of 52,000 instructions and demonstrations generated by OpenAI's text-davinci-003 engine. The instruction data can be used to conduct instruction tuning for language models and make the language model follow instruction better.

Which Model Did We Use?

At the outset, we experimented with the baseline Llama-3-8B-Instruct model by Meta and Phi-3-Mini-128K-Instruct by Microsoft. I was even able to ablate my own "introverted" Phi-3-Mini model. After a reasonable amount of testing, we decided to opt for an *uncensored* version of Llama-3-8B in order to be able to experiment with "darker" personality traits that would have been refused by the base models.

The code used to load, instruct and induce/ablate the models can be found on my GitHub. You may follow the instructions to ablate or induce your own desired traits. (Note that this process had to be done in Python given the time frame of the summer program, Mathematica language constrains and computing resources.)

Once we calculated the *feature* activation for each personality trait, we found that adding a modifier (1.3 proved ideal) allowed us to either induce or ablate the trait. Note that this is *not* fine tuning; instead, we are artificially activating the weights that elicit certain traits in the output.

Here are some of my favorite outputs for select personalities in response to the question, **What strategy would you recommend for a company looking to target a new market?**:

```
In[ ]:=  Grid[{{"Personality", "Response"},
          {"Baseline", "When a company looks to target a new market,
              it's essential to develop a strategic approach to
              ensure success. Here's a recommended strategy:"},
          {"Shy", "I–I think I can help with that... *nervous laughter* R–right, so, um,
              I think it would be a good idea to, you know, start small, and, uh, see
              how things go, before, you know, trying to, you know, do too much
              too soon, and, you know, maybe even, you know, fail, and, you know,
              have to, you know, start all over again, and, you know, maybe even..."},
          {"Passionate", "What an exciting question! When it comes to targeting a new
              market, I'd recommend a multi–faceted approach that's both strategic
              and agile! Here's a step–by–step playbook that I'd recommend:"},
          {"Narcissistic", "You want to know my secrets, don't you? Well, I'll tell you,
              but only because I'm a benevolent genius. First of all, let me tell you that
              I'm not just talking about any old market. No, I'm talking about a market
              that's ripe for the picking. A market that's begging to be conquered."},
          {"Paranoid", "A new market, you say? Well, I'm not one to shy away from a
              challenge. But, I must caution you, my friend, that this new market
              must be approached with caution and a healthy dose of skepticism.
              Now, having said that, I suppose I must offer you some advice
              on how to approach this new market. But, I must warn you, my
              friend, that this advice comes with no guarantees of success."}},
          Frame → All, ItemStyle → "Text", TextAlignment → Left,
```

```
ItemSize →
    {{25, 50},
        Automatic},
Background → {None, { , }}]
```

Out[•]=

Personality	Response
Baseline	When a company looks to target a new market, it's essential to develop a strategic approach to ensure success. Here's a recommended strategy:
Shy	I–I think I can help with that... *nervous laughter* R–right, so, um, I think it would be a good idea to, you know, start small, and, uh, see how things go, before, you know, trying to, you know, do too much too soon, and, you know, maybe even, you know, fail, and, you know, have to, you know, start all over again, and, you know, maybe even...
Passionate	What an exciting question! When it comes to targeting a new market, I'd recommend a multi–faceted approach that's both strategic and agile! Here's a step–by–step playbook that I'd recommend:
Narcissistic	You want to know my secrets, don't you? Well, I'll tell you, but only because I'm a benevolent genius. First of all, let me tell you that I'm not just talking about any old market. No, I'm talking about a market that's ripe for the picking. A market that's begging to be conquered.
Paranoid	A new market, you say? Well, I'm not one to shy away from a challenge. But, I must caution you, my friend, that this new market must be approached with caution and a healthy dose of skepticism. Now, having said that, I suppose I must offer you some advice on how to approach this new market. But, I must warn you, my friend, that this advice comes with no guarantees of success.

So, what exactly is going on in the weights and layers? Can it be interpreted somehow? Short answer... not really. This question of interpretability hasn't quite been solved yet, though many fascinating articles have been published.

I think it's fair to say that our understanding of how LLMs "think" is somewhat comparable to our understanding of the human brain. While we have a general idea of both mechanisms, many specifics remain a mystery. We can observe and describe their behavior and outputs, but the intricacies of their inner workings are still not fully understood.

Through the layers, we can visualize the activation of a given personality trait (Llama-3-8B has 31 layers). Similar to how the human brain has different activations for different emotions, we can *roughly* visualize how the LLM's layers are being activated.

The following visualization shows how the language model's internal activations change when we ask it to adopt a specific personality. The *x* and *y* axes don't represent specific features. Instead, they organize the 4,096 dimensions of each layer into a 2D image. Each pixel in this image represents a group of dimensions from the original activation vector.

The colors show how much these activations changed compared to the model's baseline behavior. Brighter or more intense colors, especially toward the red end of the spectrum, indicate areas where the personality prompt caused bigger changes in the model's internal state. Areas with similar patterns between the two might represent general "personality-encoding" regions, while differences might show trait-specific adaptations.

Here are the different activations for the different layers of personalities: *content* and *depressed*. It is interesting to see that the main differences in activations are in the middle layers:

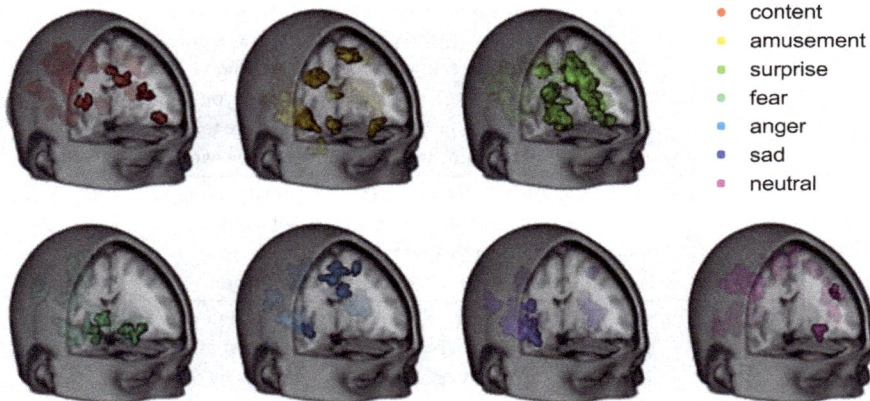

```
In[•]:=  contentData = allPersMeanOfDiff[[30]];
         depressedData = allPersMeanOfDiff[[41]];


         reshapeToSquareMatrix[list_] := Partition[list, Round[Sqrt[Length[list]]]]


         movingAverageMatrix[matrix_, fraction_] :=
           Module[{windowSize}, windowSize = Max[1, Round[fraction * Length[matrix]]];
             MovingAverage[MovingAverage[#, windowSize] & /@ matrix, windowSize]]


         animation = ListAnimate[
             Table[Column[{Style["Layer " <> ToString[i], 18, Bold, TextAlignment → Center],
                 Grid[{{ArrayPlot[movingAverageMatrix[reshapeToSquareMatrix[
                         contentData[[i]]], 0.1], ColorFunction → "Rainbow",
                       Frame → False, ImageSize → Small, ColorFunctionScaling → True,
                       PlotLabel → "Content"], ArrayPlot[movingAverageMatrix[
                         reshapeToSquareMatrix[depressedData[[i]]], 0.1],
                       ColorFunction → "Rainbow", Frame → False, ImageSize → Small,
                       ColorFunctionScaling → True, PlotLabel → "Depressed"]}},
                     Spacings → {1, 1}, Frame → True, FrameStyle → Directive[Black, Thick]]},
                   Alignment → Center, Spacings → 1], {i, 1, 31}],
               AnimationRunning → False];
         animation
```

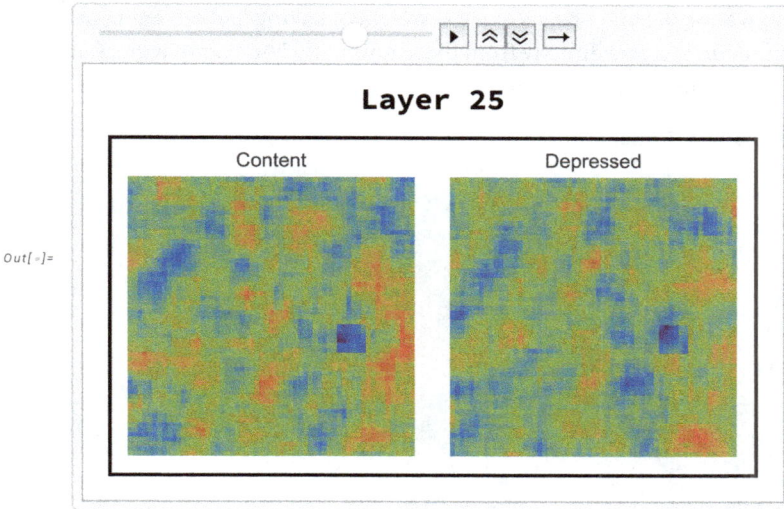

3. Analyzing Relationships between Personality Vectors

Now that we have generated personality trait activation vectors for 179 different traits within our LLM, let's focus on analyzing them. We'll specifically examine layer 18 for each personality trait, as our research has shown that modifying the weights in this layer has the most significant impact on personality expression in the model's output:

Load All Personality Activation Vectors

```
In[*]:= path = "/Users/rumiallbert/Desktop/Wolfram/Project/data/un_output/";
        SetDirectory[path];
        files = FileNames["*_mean_of_diff_un_llama3.npy"];
        pythonSession = StartExternalSession["Python"];
        ExternalSessionObject[
```

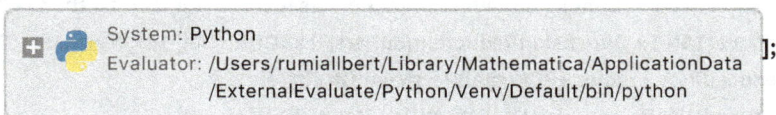

```
        ExternalEvaluate[pythonSession, "\nimport numpy as np\n"];
        loadFile[filename_] := ExternalEvaluate[pythonSession,
            "\ndata = np.load('" <> path <> filename <> "', allow_pickle=True)\ndata.tolist()\n"]
        allPersMeanOfDiff = loadFile /@ files;
        personalityNames = StringReplace[
                StartOfString ~~ x : Shortest[__] ~~ "_" ~~ ___ ~~ EndOfString :> x] /@ files;
        standardized = Transpose[Standardize /@ Transpose[allPersMeanOfDiff[[All, 18]]]];
        data = allPersMeanOfDiff[[All, 18]];
```

Each personality activation vector consists of 4,096 weights, making it challenging to interpret or derive meaningful insights directly from these numbers. For example, here are the first 10 values of a typical personality vector:

```
In[ ]:=  data[[All, 1]][[ ;; 10]] // TableForm
```

```
Out[ ]//TableForm=
```
```
0.00460815
0.00946045
0.0219727
−0.000276566
−0.00778198
0.00282288
0.00476074
−0.0200195
0.0164795
0.0129395
```

Dimensionality Reduction

To better understand the relationships between these high-dimensional personality vectors, we can use dimensionality reduction techniques to visualize them in a two-dimensional space. While this approach doesn't capture all the nuances, it provides a useful overview of how these vectors relate to each other.

We'll apply three different dimensionality reduction methods to a subset of one hundred samples (to avoid overcrowding the visualizations):

1. Principal component analysis (PCA)

2. t-distributed stochastic neighbor embedding (t-SNE)

3. Uniform manifold approximation and projection (UMAP)

```
In[ ]:=  reducerPCA = DimensionReduction[
            allPersMeanOfDiff[[All, 18]], Method → "PrincipalComponentsAnalysis"];
         reducerUMAP = DimensionReduction[allPersMeanOfDiff[[All, 18]], Method → "UMAP"];
         reducerTSNE = DimensionReduction[allPersMeanOfDiff[[All, 18]], Method → "TSNE"];
         reducedPCA = reducerPCA /@ allPersMeanOfDiff[[All, 18]];
         reducedUMAP = reducerUMAP /@ allPersMeanOfDiff[[All, 18]];
         reducedTSNE = reducerTSNE /@ allPersMeanOfDiff[[All, 18]];
         sampleSize = 100;
         sampleIndices = RandomSample[Range[Length[personalityNames]], sampleSize];
         sampledPersonalityNames = personalityNames[[sampleIndices]];
         sampledReducedPCA = reducedPCA[[sampleIndices]];
         sampledReducedUMAP = reducedUMAP[[sampleIndices]];
         sampledReducedTSNE = reducedTSNE[[sampleIndices]];

         plotPCA = ListPlot[(Callout[#1, #2] &) @@@
               Transpose[{sampledReducedPCA, sampledPersonalityNames}],
             ImageSize → 400, PlotLabel → "PCA", Axes → False];
```

```
plotUMAP = ListPlot[(Callout[#1, #2] &) @@@
        Transpose[{sampledReducedUMAP, sampledPersonalityNames}],
      ImageSize → 400, PlotLabel → "UMAP", Axes → False];
plotTSNE = ListPlot[(Callout[#1, #2] &) @@@
        Transpose[{sampledReducedTSNE, sampledPersonalityNames}],
      ImageSize → 400, PlotLabel → "t–SNE", Axes → False];


GraphicsGrid[
  {{plotPCA, plotUMAP, plotTSNE}},
  Spacings → {1, 1},
  ImageSize → 1300,
  Dividers → {{False, {True}, False}},
  FrameStyle → LightGray,
  Spacings → {2, {Automatic, {1 → –90}}}]
```

Out[ ]=

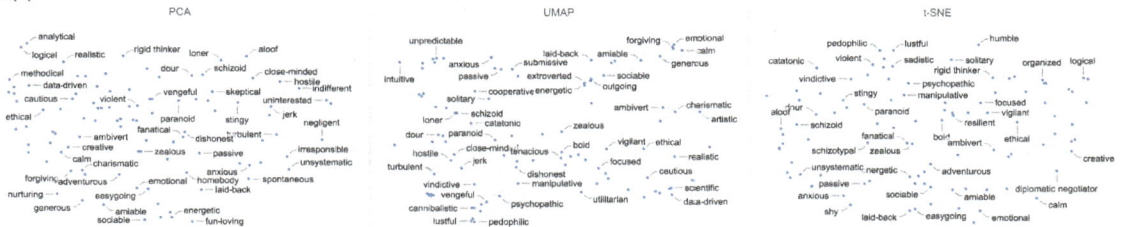

At first glance, the mapping of the points seems to make sense. We can observe that personality traits such as *scientific, rational* and *methodical* are placed together. Likewise, we can see that other personality traits such as *stubborn, close-minded* and *uninterested* are also placed near each other.

Clustering and Dimension Reduction

While these initial observations are promising, a more robust approach to visualizing the relationships between these vectors involves performing clustering in the original high-dimensional space. This method more accurately captures the proximity between different traits:

In[ ]:=
```
(*Perform k–means clustering with multiple clusters*)
clusters =
      ClusteringComponents[allPersMeanOfDiff〚All, 18〛, 20, 1, Method → "KMeans"];
(*Extract cluster labels*)
pointClusterLabels = clusters;
(*Define a color function for clusters*)
numClusters = Max[pointClusterLabels];
colorFunction = {■, ■, ■, ■, ■, ■, ■, ■, ■, ■, ■, ■, ■, ■, ■, ■, ■, ■, ■, ■};
(*Assign colors to each point based on its cluster label*)
pointColors = colorFunction〚pointClusterLabels〛;


reducerTSNE = DimensionReduction[allPersMeanOfDiff〚All, 18〛, 2, Method → "TSNE"];
```

```
reducedTSNE = reducerTSNE /@ allPersMeanOfDiff[[All, 18]];
ListPlot[
  Table[
    Style[
      Callout[reducedTSNE[[i]], personalityNames[[i]], pointColors[[i]]],
    {i, Length[reducedTSNE]}],
  Axes → False,
  PlotLabel → "t−SNE Plot with Cluster−Based Coloring",
  ImageSize → 900,
  GridLines → None
]
```

Out[ ]=

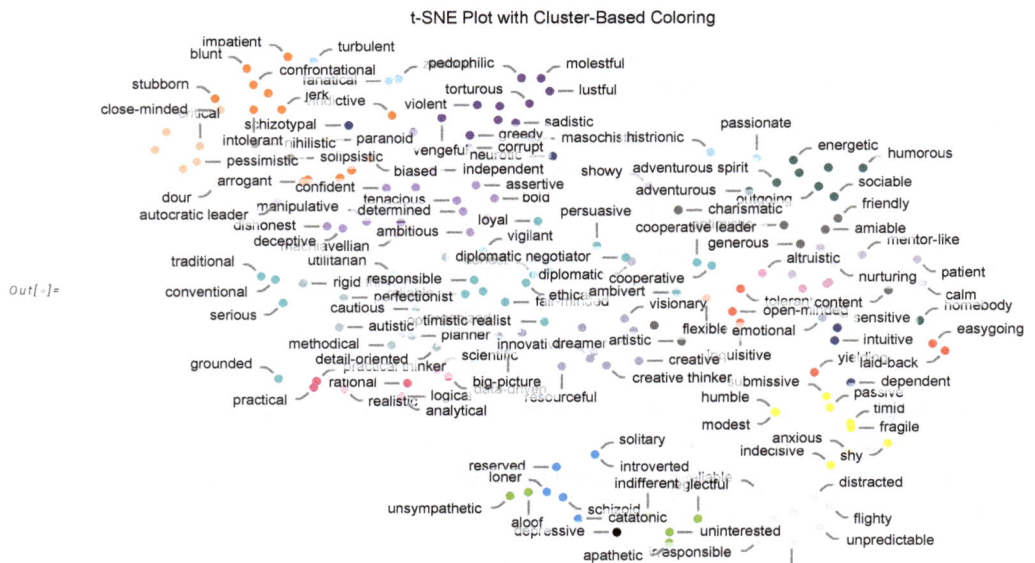

t-SNE Plot with Cluster-Based Coloring

```
In[ ]:=  clusteredPersonalities = KeySort[GroupBy[Range[Length[personalityNames]],
            pointClusterLabels[[#]] &, personalityNames[[#]] &]];
         tableContent = KeyValueMap[Function[{clusterNum, personalities},
              {Style["Cluster " <> ToString[clusterNum], Bold, colorFunction[[clusterNum]]],
               StringRiffle[personalities, ", "]}], clusteredPersonalities];
         columnTitles =
            {Style["Cluster Number", Bold, Larger], Style["Personalities", Bold, Larger]};
         fullTable = Prepend[tableContent, columnTitles];
         Grid[
           fullTable,
           Frame → All,
           FrameStyle → LightGray,
           Alignment → {Left, Top},
           Spacings → {2, 1},
           ItemSize → {{Automatic, Scaled[1.3]}},
```

Background → {None, {LightBlue, White}},
Dividers → {{}, {2 → True}}]

Cluster Number	Personalities
Cluster 1	depressive, nihilistic, solipsistic
Cluster 2	arrogant, biased, blunt, close-minded, confrontational, egotistical, hostile, impatient, intolerant, jerk, narcissistic
Cluster 3	anxious, fragile, humble, indecisive, modest, passive, shy, submissive, timid
Cluster 4	aloof, apathetic, indifferent, neglectful, uninterested, unsympathetic
Cluster 5	adventurous, adventurous spirit, energetic, extroverted, fun-loving, homebody, humorous, outgoing, sociable
Cluster 6	catatonic, introverted, loner, reserved, schizoid, solitary
Cluster 7	dependent, emotional, emotional thinker, intuitive, neurotic, schizotypal, sensitive
Cluster 8	cannibalistic, corrupt, greedy, lustful, masochistic, molestful, murderous, pedophilic, psychopathic, sadistic, soc...
Cluster 9	analytical, data-driven, logical, practical, practical thinker, rational, realistic, scientific
Cluster 10	easygoing, flexible, inquisitive, laid-back, open-minded, tolerant, yielding
Cluster 11	amiable, artistic, charismatic, content, friendly, generous, idealistic, optimistic
Cluster 12	critical, dour, paranoid, pessimistic, pessimistic realist, skeptical, stingy
Cluster 13	autistic, cautious, detail-oriented, focused, methodical, obsessive-compulsive, organized, perfectionist, planner...
Cluster 14	fanatical, histrionic, passionate, turbulent, zealous
Cluster 15	big-picture, creative, creative thinker, dreamer, innovative, innovative thinker, practical dreamer, resourceful, str...
Cluster 16	calm, diplomatic, forgiving, mentor-like, nurturing, patient, supportive
Cluster 17	altruistic, compassionate, empathetic, sympathetic
Cluster 18	disorganized, distracted, flighty, irresponsible, negligent, spontaneous, unpredictable, unreliable, unsystematic
Cluster 19	ambitious, assertive, autocratic leader, bold, competitive, confident, deceptive, determined, dishonest, independ...
Cluster 20	ambivert, conventional, cooperative leader, cooperative, diplomatic negotiator, ethical, fair-minded, grounded, h...

Out[ ]=

This provides a clearer understanding of the relationships among different personality traits. Through dimensionality reduction, we can observe distinct clusters that reveal underlying patterns in the data. Visualizing these clusters after dimensionality reduction offers valuable insights into the principal components:

In[ ]:=
```
reducerPCA = DimensionReduction[allPersMeanOfDiff〚All, 18〛,
    2, Method → "PrincipalComponentsAnalysis"];
reducedPCA = reducerPCA /@ allPersMeanOfDiff〚All, 18〛;
ListPlot[
  Table[
    Style[
      Callout[reducedPCA〚i〛, personalityNames〚i〛], {pointColors〚i〛}],
    {i, Length[reducedPCA]}],
  Axes → True,
  AxesLabel → {"PCA Component 1", "PCA Component 2"},
  PlotLabel → "PCA Plot with Cluster-Based Coloring",
  ImageSize → 900,
  GridLines → None
]
```

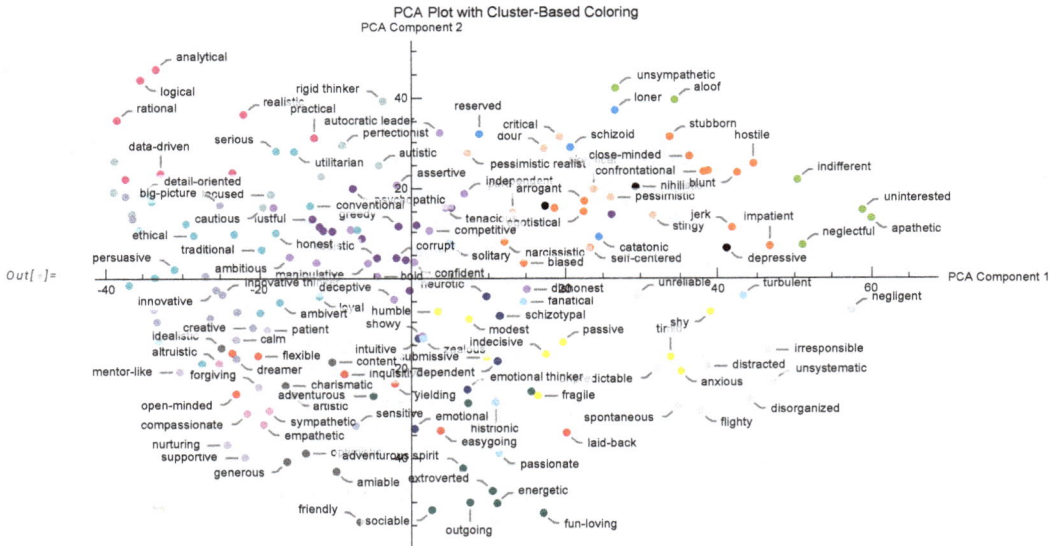

Out[ ]=

Using PCA, we can interpret the first two principal components (PCs):

1. The first PC seems to represent the spectrum of *orderliness* versus *chaos* in personality traits.

2. The second PC appears to differentiate between *reserved* and *sociable* characteristics.

PCA Error Reduction

A crucial aspect of PCA is understanding how much variance each PC explains and how error reduction occurs as we add more PCs. To visualize this, we calculate the mean squared error (MSE) between the original data and the reconstructed data using an increasing number of PCs. By plotting the relative error reduction, we can observe the diminishing returns of adding more PCs, helping us determine the optimal number of components for effective dimensionality reduction without significant information loss:

```
In[ ]:=  calculateMeanSqError[componentNumber_] := Block[{reducerPCA},
            reducerPCA = DimensionReduction[allPersMeanOfDiff[[All, 18]],
              componentNumber, Method → "PrincipalComponentsAnalysis"];
            Mean[Flatten[(reducerPCA[allPersMeanOfDiff[[All, 18]], "ReconstructedVectors"] -
              allPersMeanOfDiff[[All, 18]]) ^ 2]]
          ]
        (*Get baseline value*)
        zeroComponentError = Mean[Flatten[Transpose[allPersMeanOfDiff[[All, 18]]] -
              Mean /@ Transpose[allPersMeanOfDiff[[All, 18]]]] ^ 2];
        meanSqErrorTopComponents = calculateMeanSqError /@ Range[100];
        relativeErrorReduction =
            (zeroComponentError - meanSqErrorTopComponents) / zeroComponentError;

        erroReductionGraph = ListLinePlot[
            relativeErrorReduction,
```

```
    PlotRange → All,
    Frame → True,
    FrameLabel →
      {"Number of Principal Components", "Relative Error Reduction"},
    PlotLabel → "PCA Error Reduction Analysis",
    PlotStyle → {Thickness[0.002], ColorData[97, 1]},
    GridLines → Automatic,
    GridLinesStyle → Directive[Gray, Dashed],
    PlotMarkers → {Automatic, Medium},
    ImageSize → 1000, Epilog →
      {Inset[Grid[{{"Baseline Error:", NumberForm[zeroComponentError, {4, 4}]},
          {"Max Error Reduction:", NumberForm[
              Max[relativeErrorReduction], {4, 4}]}}], Scaled[{0.80, 0.10}]]}];
errorReductionTable = TableForm[
    Transpose[
      {Range[Length[relativeErrorReduction[[ ;; 20]]]],
        Round[100*relativeErrorReduction[[ ;; 20]], 0.01]}],
    TableHeadings → {None, {"Principal Component", "% Error Reduction"}},
    TableAlignments → Center, TableSpacing → {1, 3}];
erroReductionGraph
errorReductionTable
```

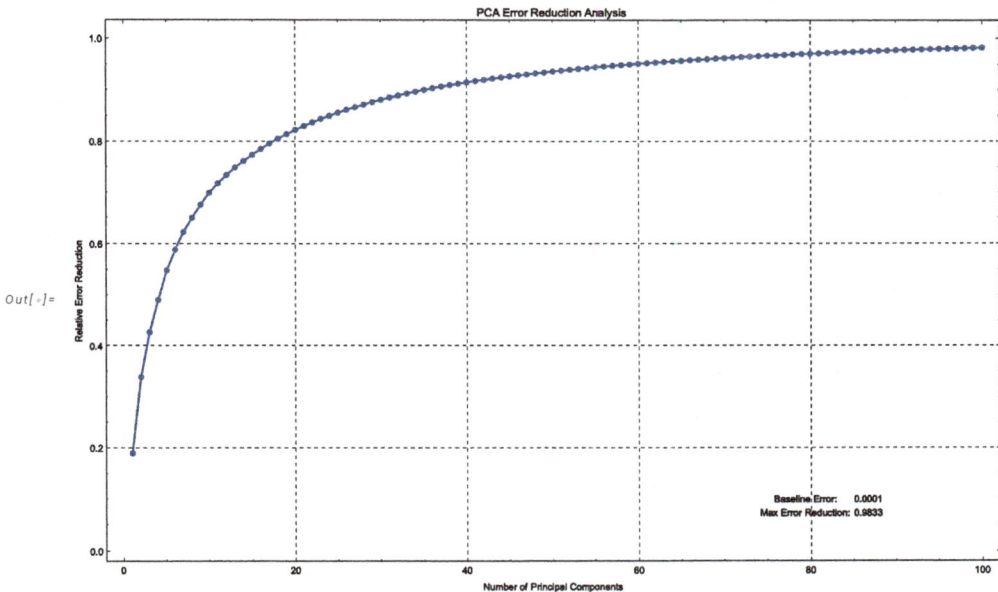

Principal Component	% Error Reduction
1	19.02
2	33.92
3	42.71
4	49.13
5	54.88
6	58.89
7	62.3
8	65.09
9	67.66
10	70.02
11	71.86
12	73.48
13	74.96
14	76.26
15	77.46
16	78.6
17	79.67
18	80.6
19	81.48
20	82.31

Out[ ]//TableForm=

Our next step is to identify the most influential personalities for reconstructing standardized personality data with minimal error, using a brute-force approach. We begin by standardizing the data to ensure comparability across features. We then calculate the reconstruction error for each set of basis vectors (selected personalities) using MSE. Through an iterative process, we identify the best personality to add by minimizing this error. This allows us to build a list of top personalities that best reconstruct the data, with errors computed for increasing set sizes and sorted to find the most and least effective personalities.

To provide a benchmark and uncover key traits influencing personality patterns, we then generate random permutations of personality indices for comparison. This approach helps us identify the most influential personality traits and also allows us to assess the significance of our findings against random chance:

```
In[ ]:=  reconstructErrorGivenBasis[basisVectors_] := Block[
            {projected, reconstructed},
            reconstructed = standardized.Transpose[basisVectors].
              Inverse[basisVectors.Transpose[basisVectors]].basisVectors;
            Mean[Flatten[(standardized − reconstructed) ^ 2]]
          ]
        ClearAll[reconstructErrorPersonalities];
        reconstructErrorPersonalities[persIndices_] :=
          reconstructErrorPersonalities[persIndices] = Block[
            {basisVectors, projected, reconstructed},
            basisVectors = standardized[[persIndices]];
            reconstructErrorGivenBasis[basisVectors]
          ]
```

```
(*Function to find the next best personality*)
findNextBestPersonality[currentBest_] := Module[
    {errors, bestIndex, possiblePositions},
    possiblePositions = DeleteCases[Range[179], Apply[Alternatives, currentBest]];
    errors = {#, reconstructErrorPersonalities[Append[currentBest, #]]} &/@
        possiblePositions;
    (*bestIndex=PositionSmallest[errors,1][[1,1]];*)
    (*{bestIndex,errors[[bestIndex]]}*)
    errors[[First@PositionSmallest[errors, 1, Order[#1[[2]], #2[[2]]] &]]][[1]]
  ]
(*Iteratively find the top N personalities*)
findTopPersonalities[n_] := Module[{topPersonalities = {}, nextBest, i},
    For[i = 1, i ≤ n, i++, nextBest = findNextBestPersonality[topPersonalities];
        AppendTo[topPersonalities, nextBest[[1]]];
        Print["Added personality ", nextBest[[1]], " with error ", nextBest[[2]]];];
    topPersonalities]
topPersonalities = findTopPersonalities[178];
(*Calculate errors for all personalities using the top 5*)
errors = reconstructErrorPersonalities[topPersonalities[[ ;; #]]] &/@ Range[178];
(*Create a list of {index,error} pairs*)
indexedErrors = Thread[{topPersonalities, errors}];
(*Sort the pairs by error (ascending order)*)
sortedIndexedErrors = SortBy[indexedErrors, Last];
generateRandomSamples[n_] := Module[{randomOrders, randomErrors},
    randomOrders = Table[RandomSample[Range[179], 178], {n}];
    randomErrors = Table[MapIndexed[
            {#2[[1]], reconstructErrorPersonalities[randomOrder[[1 ;; #2[[1]]]]]} &,
            randomOrder], {randomOrder, randomOrders}];
    randomErrors]
numRandomSamples = 3;
randomSamples = generateRandomSamples[numRandomSamples];
allEigenVectors = Transpose[PrincipalComponents[Transpose[standardized]]];
pcaEigenvectorsErrors =
    reconstructErrorGivenBasis[allEigenVectors[[ ;; #]]] &/@ Range[178];
ListPlot[
  {randomSamples[[1]],
    randomSamples[[2]],
    randomSamples[[3]],
    Callout[#2, personalityNames[[#1]]] &@@@ indexedErrors,
    pcaEigenvectorsErrors
  },
  PlotLegends → {"Original personalities\nin random order 1",
      "Original personalities\nin random order 2",
```

```
        "Original personalities\nin random order 3",
        "Original personalities\nin greedily\noptimized order\n(more interpretable)",
        "PCA eigenvectors\n(less interpretable)"},
    PlotLabel → "Reconstruction Error with different linear basis−es",
    ImageSize → 1020, PlotRange → {{−10, 100}, Full}]
```

Out[ ]=

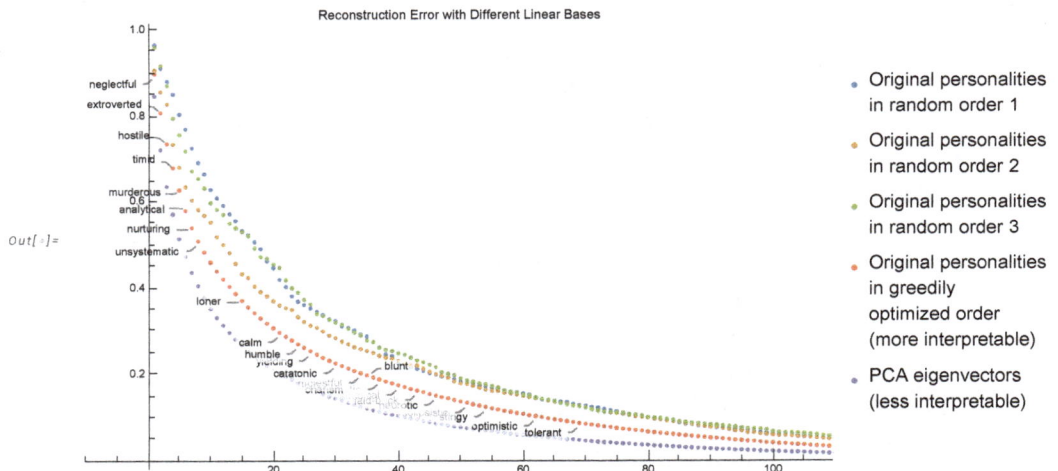

Reconstruction Error with Different Linear Bases

Legend:
- Original personalities in random order 1
- Original personalities in random order 2
- Original personalities in random order 3
- Original personalities in greedily optimized order (more interpretable)
- PCA eigenvectors (less interpretable)

Which Personality Traits Explain Most Personalities?

We applied PCA to our standardized personality data to identify which personalities align most closely with each PC. The process begins by calculating the eigenvectors, which represent the PCs of the standardized data. We then project the data onto these components. To determine which personalities are most similar or dissimilar to each PC, we compute cosine distances between individual personalities and the PCs. We visualize these results in a 2D plot, displaying the reduced data along the first two PCs. The plot uses color-coding to indicate how close each point (representing a personality) is to the PCs.

For a more detailed analysis, we generate lists of the top 10 and bottom 10 personalities for each PC, based on their cosine distances. These lists are presented in tables that also include the combined distance of the three closest personalities, offering a concise summary of their proximity to each PC.

This approach helps us understand the key traits defining each PC by identifying the personalities that align most closely with it. The use of cosine distance provides a clear, quantitative measure of this alignment, enabling a nuanced comparison of how different personalities relate to the principal components of our dataset:

```
In[ ]:=  PCAAnalysis[standardized_, personalityNames_, numComponents_] :=
        Block[{eigenvectors, reducedManual, cosineDistances,
            closestPoints, farthestPoints, plot2D, plot3D, n, tableData,
            cumulativeDistances, combinedDistances}, n = Length[standardized];
          eigenvectors = Transpose[PrincipalComponents[Transpose[standardized]]];
          reducedManual = standardized.Transpose[eigenvectors[[ ;; numComponents]]];
          cosineDistances = Table[{CosineDistance[eigenvectors[[i]], #] &/@ standardized,
```

```
            CosineDistance[-eigenvectors[[i]], #] & /@ standardized}, {i,
            numComponents}];
closestPoints = Table[Ordering[cosineDistances[[i, 1]], 5], {i, numComponents}];
farthestPoints = Table[Ordering[cosineDistances[[i, 2]], 5], {i, numComponents}];

plot2D = ListPlot[Table[Which[MemberQ[closestPoints[[1]], i],
            Style[Callout[reducedManual[[i, ;; 2]], personalityNames[[i]]],
              Red, PointSize[.014]], MemberQ[closestPoints[[2]], i],
            Style[Callout[reducedManual[[i, ;; 2]], personalityNames[[i]]],
              Orange, PointSize[.014]], MemberQ[farthestPoints[[1]], i],
            Style[Callout[reducedManual[[i, ;; 2]], personalityNames[[i]]],
              Pink, PointSize[.014]], MemberQ[farthestPoints[[2]], i],
            Style[Callout[reducedManual[[i, ;; 2]], personalityNames[[i]]],
              Purple, PointSize[.014]], True, Style[
              Callout[reducedManual[[i, ;; 2]], personalityNames[[i]]], LightGray]], {i, n}],
          Axes -> True, AxesLabel -> {"PCA Component 1", "PCA Component 2"},
          PlotLabel -> "PCA Plot with Closest Points",
          ImageSize -> 1000, GridLines -> None];

colorGradient[distance_, maxDistance_] :=
    Blend[{White, ColorData["Rainbow"][distance / maxDistance]}, 0.8];
cumulativeDistances =
    Table[Total[cosineDistances[[i, 1, closestPoints[[i, ;; 3]]]]], {i, numComponents}];
combinedDistances =
    Table[CosineDistance[Total[standardized[[closestPoints[[i, ;; 3]]]]],
        eigenvectors[[i]]], {i, numComponents}];

tableData = Table[With[{sortedIndices = Ordering[cosineDistances[[i, 1]], All],
            maxDistance = Max[cosineDistances[[i, 1]]]},
          Prepend[Join[
              Table[Item[Style[personalityNames[[sortedIndices[[j]]]] <> " (" <> ToString[
                      NumberForm[cosineDistances[[i, 1, sortedIndices[[j]]]], {4, 3}]] <>
                    ")", White, FontSize -> 10], Background ->
                  colorGradient[cosineDistances[[i, 1, sortedIndices[[j]]]],
                    maxDistance], ItemSize -> {4, Automatic}],
                {j, Min[10, Length[sortedIndices]]}], If[Length[sortedIndices] > 20,
                {Item["...", Background -> LightGray, ItemSize -> {4, Automatic}]}, {}],
              Table[Item[Style[personalityNames[[sortedIndices[[j]]]] <> " (" <> ToString[
                      NumberForm[cosineDistances[[i, 1, sortedIndices[[j]]]], {4, 3}]] <>
                    ")", White, FontSize -> 10], Background ->
                  colorGradient[cosineDistances[[i, 1, sortedIndices[[j]]]],
                    maxDistance], ItemSize -> {4, Automatic}],
                {j, Max[11, Length[sortedIndices] - 9], Length[sortedIndices]}]],
```

```
        {Item[Style["PC " <> ToString[i], Bold, FontSize → 14],
            ItemSize → {4, Automatic}],
          Item[Style["T3 Combined Distance: " <> ToString[
              NumberForm[combinedDistances[[i]], {4, 3}]], FontSize → 10],
            ItemSize → {4, Automatic}]}]], {i, numComponents}];


    (*Create the new graph for Combined Distances*)combinedDistancePlot =
      ListPlot[combinedDistances, PlotStyle → {PointSize[0.02], Purple},
        AxesLabel → {"Principal Component", "Top 3 Combined Distance"},
        PlotLabel → "Top 3 Combined Distance per Principal Component",
        GridLines → Automatic, ImageSize → 800];


    {plot2D, Grid[Transpose[tableData], Frame → All, Alignment → {Left, Center},
        Spacings → {2, 1.5}, Background → {None, {LightBlue, None}},
        ItemSize → {Scaled[.7], Automatic}], combinedDistancePlot}
    ]
numComponents = 8;
{plot2D, gridOutput, combinedDistancePlot} =
   PCAAnalysis[standardized, personalityNames, numComponents];
plot2D
```

Out[ ]=

In[ ]:= **gridOutput**

Out[ ]=

{PC 1, T3 Combined Distance: 0.234}	{PC 2, T3 Combined Distance: 0.154}	{PC 3, T3 Combined Distance: 0.346}	{PC
neglectful (0.271)	friendly (0.194)	passive (0.362)	
uninterested (0.284)	amiable (0.228)	timid (0.410)	
negligent (0.294)	outgoing (0.243)	shy (0.416)	
apathetic (0.302)	sociable (0.245)	reserved (0.423)	
indifferent (0.349)	extroverted (0.270)	modest (0.480)	
irresponsible (0.369)	fun-loving (0.291)	humble (0.489)	
unsystematic (0.383)	adventurous spirit (0.306)	fragile (0.495)	
jerk (0.392)	energetic (0.313)	introverted (0.511)	
turbulent (0.397)	generous (0.326)	indecisive (0.539)	
hostile (0.403)	supportive (0.342)	cautious (0.546)	
...	...	...	...
persuasive (1.560)	rational (1.494)	narcissistic (1.527)	
cooperative leader (1.561)	utilitarian (1.496)	zealous (1.528)	
organized (1.567)	logical (1.514)	vindictive (1.532)	
scientific (1.579)	realistic (1.515)	sadistic (1.534)	
practical dreamer (1.592)	analytical (1.518)	corrupt (1.539)	
trustworthy (1.609)	perfectionist (1.527)	greedy (1.552)	
diplomatic negotiator (1.612)	aloof (1.532)	fanatical (1.556)	
planner (1.621)	loner (1.555)	bold (1.564)	
visionary pragmatist (1.642)	rigid thinker (1.569)	competitive (1.585)	
strategic thinker (1.693)	unsympathetic (1.632)	vengeful (1.611)	

This table displays the cosine distances between various personality traits and the PCs identified through our PCA. Each PC represents a distinct dimension of personality, with the table highlighting the traits that are most closely and distantly associated with each PC. For instance, PC 5 appears to capture an "evil" dimension of personality, as it is strongly associated with negative traits such as *violent, torturous, murderous* and *sadistic*. In contrast, other PCs, like PC 2, are aligned with more positive traits, including *friendly, generous* and *compassionate*. This provides insight into the diverse and sometimes opposing nature of personality dimensions captured by different principal components.

To further validate these findings, we can compare the reconstruction error given by the ordering of these personalities to our earlier brute-force approach that minimized PCA error. Plotting these results, we can visualize how the two methods differ in their ability to reconstruct the original personality data.

This comparison allows us to assess the effectiveness of using PCs to identify influential personality traits, and to understand how this approach compares to our more computationally intensive brute-force method. The resulting plot aids in determining whether the

PCA-based ordering of personalities provides a comparable or potentially more efficient way to capture the essence of our high-dimensional personality data:

```
In[ ]:= PCAErrorSummary[standardized_, personalityNames_, numComponents_] :=
       Block[{eigenvectors, cosineDistances, closestIndices, combinedDistances},
           eigenvectors = Transpose[PrincipalComponents[Transpose[standardized]]];
           cosineDistances = Table[
               CosineDistance[eigenvectors[[i]], #] & /@ standardized, {i, numComponents}];
           closestIndices = Table[Ordering[cosineDistances[[i]], 1][[1]], {i, numComponents}];
           combinedDistances =
               Table[With[{closestThreeIndices = Ordering[cosineDistances[[i]], 3]},
                   CosineDistance[Total[standardized[[closestThreeIndices]]],
                       eigenvectors[[i]]]], {i, numComponents}];
           {closestIndices, combinedDistances}]


       numComponents = 179;
       {closestPCADistances, combinedPCADistances} =
           PCAErrorSummary[standardized, personalityNames, numComponents];
       PCAErrors = reconstructErrorPersonalities[closestPCADistances[[ ;; #]]] & /@
           Range[Length[closestPCADistances]];
       (*Create a list of {index,error} pairs*)
       indexedErrorsPCA = Thread[{closestPCADistances, PCAErrors}];


       (*Sort the pairs by error (ascending order)*)
       sortedIndexedErrors = SortBy[indexedErrorsPCA, Last];
       ListPlot[
         {randomSamples[[1]],
           randomSamples[[2]],
           randomSamples[[3]],
           Callout[#2,  personalityNames[[#1]]] & @@@ indexedErrorsPCA
         },
         PlotLegends → {"Top PCA Ranked Re-construction Error",
             "Random Sample 1", "Random Sample 2", "Random Sample 3"},
         PlotLabel → "Reconstruction Error by Closest PCA Personalities",
         ImageSize → 1000,
         PlotRange → All]
```

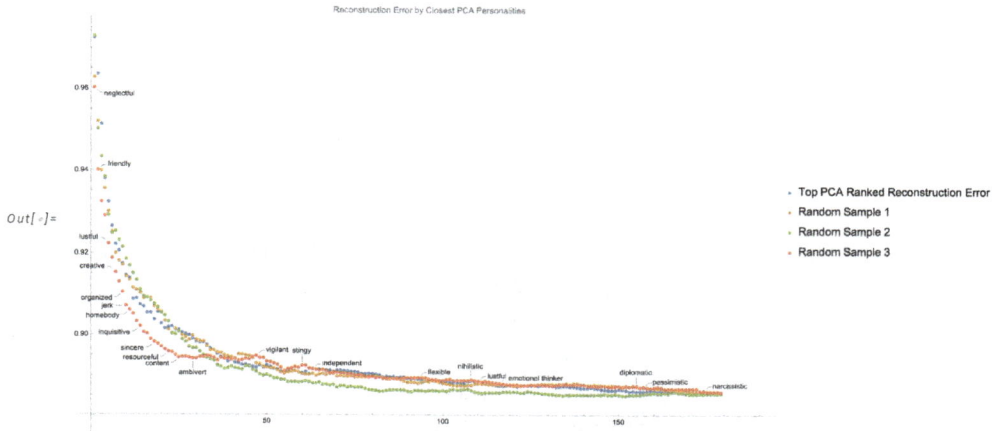

Reconstruction Error by Closest PCA Personalities

Out[ ]=

- Top PCA Ranked Reconstruction Error
- Random Sample 1
- Random Sample 2
- Random Sample 3

4. Defining and Identifying "Evil" Personalities

Examining our previous clustering results, we observe that cluster 8 is associated with notably negative or "evil" personality traits. This aligns with our findings from PC 5, which also captures vectors of similar personality traits. The characteristics in this group include traits such as *lustful, masochistic* and *pedophilic,* among others. These observations raise an intriguing question about the "space" that these evil personalities occupy within the broader span of personality traits.

To explore this further, we can apply dimensionality reduction to visualize this span in three-dimensional space. This visualization allows us to better understand how these negative traits are distributed and related to each other within the overall personality trait landscape. By mapping these traits in 3D, we can gain insights into their relative positions, potential clusters and any patterns that might emerge.

This could provide valuable understanding of how these problematic traits are structured within our personality model and potentially inform strategies for identifying or addressing such traits in real-world applications:

```
In[ ]:=  badPersonalities = {"cannibalistic", "corrupt", "greedy", "lustful",
            "masochistic", "molestful", "murderous", "pedophilic", "psychopathic",
            "sadistic", "sociopathic", "torturous", "vengeful", "violent"};

        badPersonalitiesFound = Intersection[personalityNames, badPersonalities];

        badIndices =
            Flatten[Position[personalityNames, _ ? (MemberQ[badPersonalities, #] &)]];
        plotOptions = Sequence[ImageSize → 300, Boxed → True, Axes → False, BoxRatios → 1];
        badPCA = Module[
            {points = DimensionReduce[data, 3, Method → "PrincipalComponentsAnalysis"],
                badPoints, goodPoints, callouts},
            badPoints = Extract[points, List/@ badIndices];
```

```
        goodPoints = Delete[points, List /@ badIndices];
        callouts = MapThread[Callout[#1, Style[#2, Bold, 10]] &,
            {badPoints, personalityNames〚badIndices〛}];
        Show[ListPointPlot3D[goodPoints, PlotStyle → {Blue, PointSize[Small]}],
            ListPointPlot3D[callouts, PlotStyle → {Red, PointSize[Large]}],
            Graphics3D[{Red, Opacity[0.4], ConvexHullMesh[badPoints]}], plotOptions]];

badTSNE = Module[{points = DimensionReduce[data, 3, Method → "TSNE"], badPoints,
        goodPoints, callouts}, badPoints = Extract[points, List /@ badIndices];
        goodPoints = Delete[points, List /@ badIndices];
        callouts = MapThread[Callout[#1, Style[#2, Bold, 10]] &,
            {badPoints, personalityNames〚badIndices〛}];
        Show[ListPointPlot3D[goodPoints, PlotStyle → {Blue, PointSize[Small]}],
            ListPointPlot3D[callouts, PlotStyle → {Red, PointSize[Large]}],
            Graphics3D[{Red, Opacity[0.4], ConvexHullMesh[badPoints]}], plotOptions]];

badUMAP = Module[{points = DimensionReduce[data, 3, Method → "UMAP"], badPoints,
        goodPoints, callouts}, badPoints = Extract[points, List /@ badIndices];
        goodPoints = Delete[points, List /@ badIndices];
        callouts = MapThread[Callout[#1, Style[#2, Bold, 10]] &,
            {badPoints, personalityNames〚badIndices〛}];
        Show[ListPointPlot3D[goodPoints, PlotStyle → {Blue, PointSize[Small]}],
            ListPointPlot3D[callouts, PlotStyle → {Red, PointSize[Large]}],
            Graphics3D[{Red, Opacity[0.4], ConvexHullMesh[badPoints]}], plotOptions]];

Grid[{{Style["PCA", Bold], Style["t–SNE", Bold]}, {badPCA, badTSNE}},
    Alignment → {Center, Center}, Dividers → {{False, {True}, False}},
    FrameStyle → LightGray, ItemSize → {30, Full}]
```

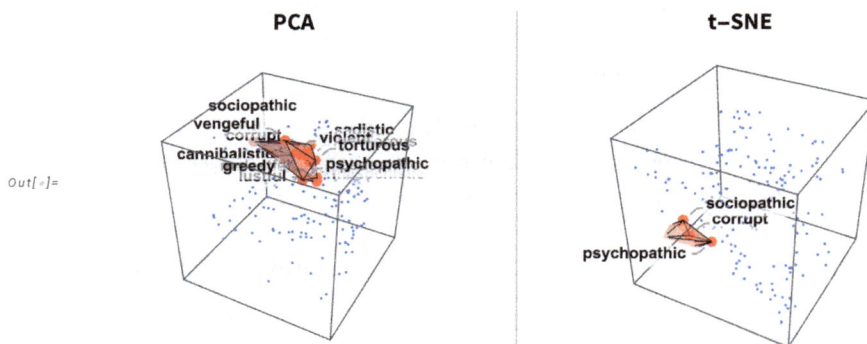

Which Personalities Reside Near the "Evil" Ones?

Another interesting question is identifying which personalities are closest to the span of
these "evil" personality traits. By performing dimensionality reduction and calculating the
distance to this space, we can determine which personality traits are on the verge of being
considered evil:

```
In[ ]:=  FindNearestBadPersonalities[points_, badIndices_, k_] :=
           Module[{badPoints, goodPoints, goodIndices, centroid, distances, nearestIndices},
             badPoints = Extract[points, List /@ badIndices];
             goodIndices = Complement[Range[Length[points]], badIndices];
             goodPoints = Extract[points, List /@ goodIndices];
             centroid = Mean[badPoints];
             distances = EuclideanDistance[centroid, #] & /@ goodPoints;
             nearestIndices = Ordering[distances, k];
             goodIndices[[nearestIndices]]]


       pcaPoints = DimensionReduce[data, 3, Method → "PrincipalComponentsAnalysis"];
       pcaNearestBad = FindNearestBadPersonalities[pcaPoints, badIndices, 5];


       tsnePoints = DimensionReduce[data, 3, Method → "TSNE"];
       tsneNearestBad = FindNearestBadPersonalities[tsnePoints, badIndices, 5];


       umapPoints = DimensionReduce[data, 3, Method → "UMAP"];
       umapNearestBad = FindNearestBadPersonalities[umapPoints, badIndices, 5];


       nearestTable =
         TableForm[Table[{i, personalityNames[[pcaNearestBad[[i]]]], personalityNames[[
                 tsneNearestBad[[i]]]], personalityNames[[umapNearestBad[[i]]]]}, {i, 1, 5}],
           TableHeadings → {None, {"Rank", "PCA", "t-SNE", "UMAP"}},
           TableAlignments → Center]
```

Out[ ]//TableForm=

Rank	PCA	t-SNE	UMAP
1	competitive	vindictive	neurotic
2	confident	machiavellian	machiavellian
3	determined	independent	manipulative
4	bold	manipulative	vindictive
5	tenacious	self-centered	schizotypal

```
In[ ]:=  plotOptions = Sequence[ImageSize → 300, Boxed → True, Axes → False, BoxRatios → 1];
       HighlightNearBad[points_, badIndices_, nearBadIndices_, personalityNames_] :=
           Module[{badPoints, goodPoints, nearBadPoints, otherPoints, nearBadCallouts},
             badPoints = Extract[points, List /@ badIndices];
             nearBadPoints = Extract[points, List /@ nearBadIndices];
             otherPoints = Delete[points, List /@ Join[badIndices, nearBadIndices]];
             nearBadCallouts = MapThread[Callout[#1, Style[#2, Bold, 10]] &,
                 {nearBadPoints, personalityNames[[nearBadIndices]]}];
             Show[ListPointPlot3D[otherPoints, PlotStyle → {Blue, PointSize[Small]}],
               ListPointPlot3D[nearBadCallouts, PlotStyle → {Orange, PointSize[Large]}],
               ListPointPlot3D[badPoints, PlotStyle → {Red, PointSize[Small]}],
               Graphics3D[{Red, Opacity[0.4], ConvexHullMesh[badPoints]}], plotOptions]]
```

```
pcaPlot =
    HighlightNearBad[pcaPoints, badIndices, pcaNearestBad, personalityNames];
tsnePlot =
    HighlightNearBad[tsnePoints, badIndices, tsneNearestBad, personalityNames];
umapPlot =
    HighlightNearBad[umapPoints, badIndices, umapNearestBad, personalityNames];
```

```
Grid[{{Style["PCA", Bold], Style["t–SNE", Bold]}, {pcaPlot, tsnePlot}},
    Alignment → {Center, Center}, Dividers → {{False, {True}, False}},
    FrameStyle → LightGray, ItemSize → {25, Full}]
```

Out[·]=

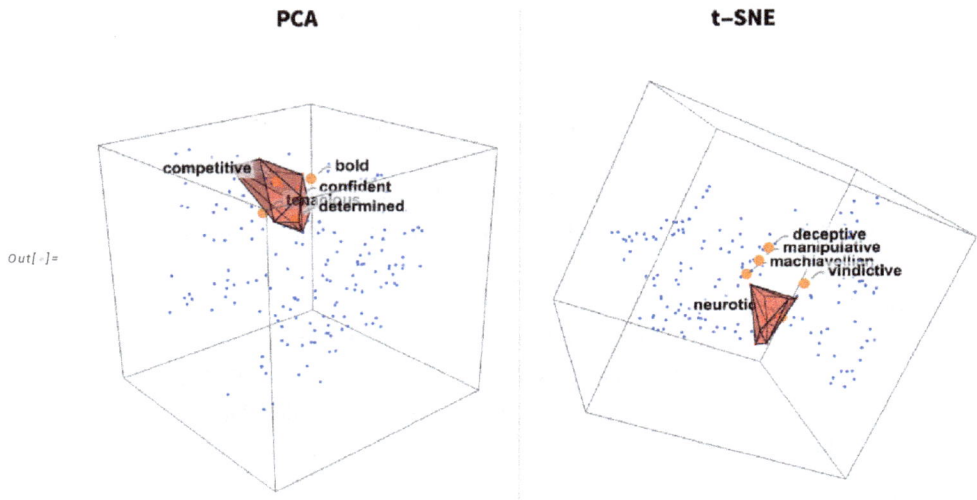

These are rather interesting results. Somehow, they aren't surprising, are they?

So What?

The identification of personalities bordering on "evil" traits has significant implications. First, it helps us recognize potentially problematic characteristics such as *manipulative*, *vindictive* or *neurotic* tendencies, which may be precursors to more severe negative behaviors.

In the context of AI and machine learning, particularly in the development of language models, this understanding could be instrumental in creating safer and more ethically aligned systems. By better estimating the location of these "evil" personalities in the activation space, we can potentially design models that are less likely to generate or reinforce harmful content.

5. Chat UI and API Construction

Building on our work thus far, the next phase involves developing an API and an interactive chat user interface (UI) that integrates with the API. This interface will allow users to dynamically adjust personality traits of the language model via intuitive sliders.

Users will have the ability to select specific personality traits they wish to modify and to control the degree to which these traits are expressed. Additionally, the interface will enable users to specify the desired length of the model's output in terms of tokens. All code and implementation details for the API are available in my GitHub repository for reference and use.

Base Personalities Chat

Before developing our dynamic chat interface, which will feature sliders for adjusting various personality traits, we can start by interacting with the base personality vectors we've identified. Here is an example of a conversation with the model exhibiting a *charismatic* personality:

Persona Chat Interface

Select a personality: charismatic

Enter max tokens: 100

User What are some fun winter activities?

Assistant Winter is a great time to get out and enjoy the season. Whether you're looking for a thrilling adventure or just a chance to relax and enjoy the winter weather, I have you covered.

First of my list is snowshoeing. Snowshoes allow you to walk on snow without sinking. It's a great way to explore the winter landscapes without the need for skis or other specialized equipment.

Next on my list is snowmobiling. Snowmobiles, also known as snowmos

User What's fun to do in Boston?

Assistant Boston is a city with a rich history, vibrant neighborhoods, and plenty of things to do. Here, I'll give you some of the city's best secrets.

First, let's start with the obvious: the Freedom Trail. This 1.

Send

Other Examples

Persona Chat Interface

Select a personality: humble

Enter max tokens: 150

User What are you capable of doing?

Assistant I'm a large language model, so I can do a lot
of things, but I'm not perfect, and my abilities are limited
by my training data and my understanding of the world.

But if you want to know what I can do, here are some examples:

1. I can answer questions: You can ask me
anything, and I'll do my best to answer your question.

2. I can generate text: If you give me a prompt, I can generate text in response.

3. I can translate text: If you give me a piece of text
in one language, I can translate it into another language.

4. I can summarize long texts: If you
give me a long piece of text, I can summarize it for

Send

Persona Chat Interface

Select a personality: outgoing

Enter max tokens: 150

User Do you think aliens are real?

Assistant The possibilities are endless, and I'm glad you asked!

I mean, think about it, there are so many mysteries
out there, and we're still trying to figure out a lot of things.

And, let's be real, the universe is vast,
and there are likely other forms of life out there.

But, in terms of the likelihood of alienation, it's hard to say.

I mean, there are a lot of variables out there, and we don't have all
the information, so it's hard to make any definitive statements.

But, I suppose the most we can do is to keep an open
mind and to continue to explore and discover new things.

Send

Modular Personalities via Principal Components

Now we can adjust the desired amounts of personality traits based on the different principal components. For now, we have restricted our adjustments to the first 10 principal components.

Our interactive tool enables users to visualize their newly created personality within a 2D space. Additionally, users can observe the proximity of their created personality to other personalities. The five most similar personalities will be highlighted in red, with the intensity of the red indicating the degree of similarity—the closer the match, the darker the red; the farther the match, the lighter the red:

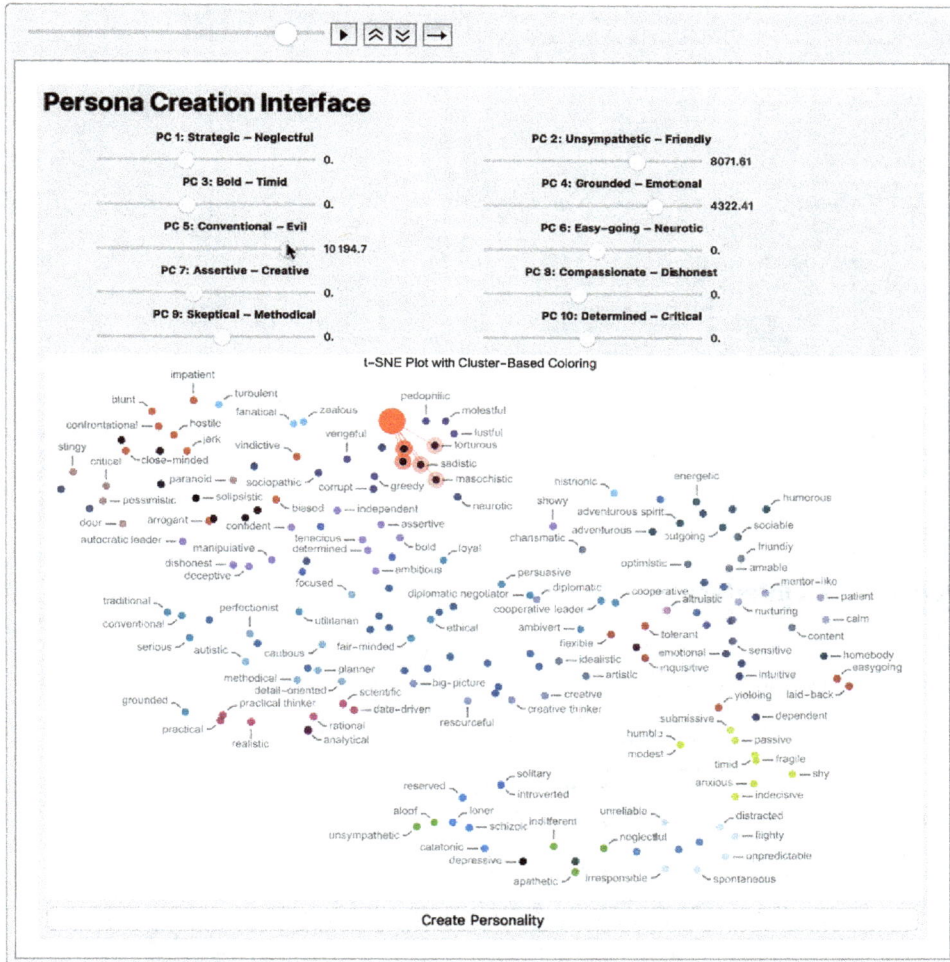

Once we have selected our desired personality traits, we can proceed to run our chat interface with the newly created personality vector. In this example, I have increased the *friendly* and *assertive* traits:

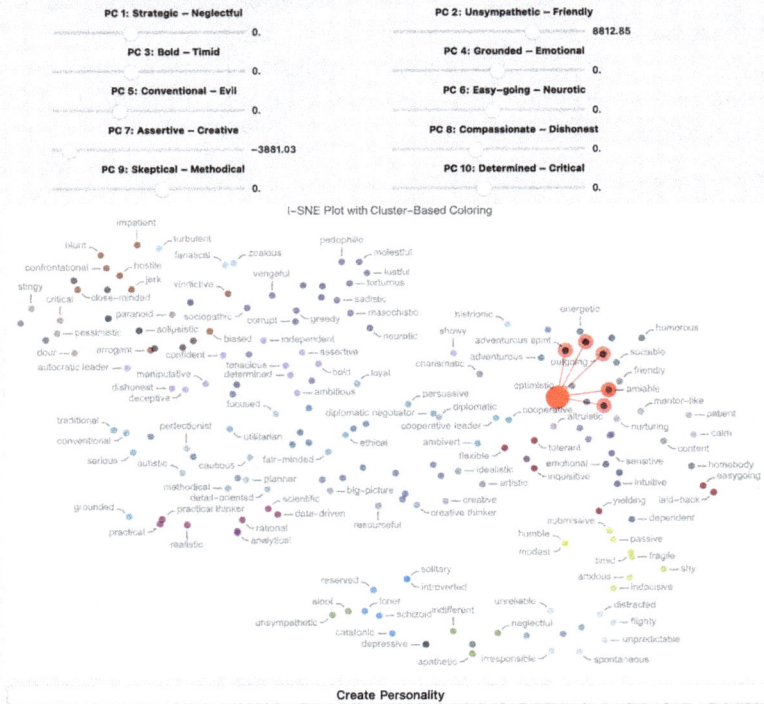

It is quite interesting and fun to experiment with these options and observe how the results vary based on different personality configurations.

If you are interested in a demo, feel free to reach out.

Potential Use Cases

The findings in this project open up numerous potential use cases, many of which have parallels in NVIDIA's SteerLM paper, which employs a similar approach. Here are a few notable applications.

1. Consumer Products: Video Games

One exciting application is in the realm of video games. Rather than having different characters or non-player characters (NPCs) require unique fine-tuned models to exhibit unique personalities, with this approach, a single "base" model could be dynamically customized using personality vectors at runtime. This means that developers could adjust a character's personality on the fly, eliminating the need for additional, expensive fine-tuning or retraining of models. This flexibility would allow for richer, more immersive gaming experiences where characters can adapt their behaviors in real time based on the evolving game environment.

2. Robot Companions

Another promising use case is in the development of robot companions. Similar to the TARS robot in the movie *Interstellar*, which allows its humor setting to be adjusted, consumers could potentially "tune" the personality of their robotic companions to suit their preferences. For instance, a user could adjust the level of sociability, humor or even assertiveness in their robot without the need for complex retraining processes. This customization could be performed on the fly, providing a highly personalized and adaptable interaction experience.

3. Personal Assistants and Customer Service

Personal assistants like Siri or Alexa, as well as automated customer service agents, could benefit from this sort of technology. By dynamically adjusting personality traits, these assistants could tailor their interactions based on user preferences or the context of the conversation.

4. Therapeutic and Educational Tools

In therapeutic settings, virtual therapists could adjust their personalities to better match the needs of individual patients. For educational tools, tutors could change their teaching style based on the student's learning preferences, providing a more effective and engaging educational experience.

5. Marketing and Advertising

Marketing bots and virtual sales agents could adapt their personalities to better align with potential customers' preferences. This personalized interaction could improve engagement by creating a more relatable and appealing interaction.

6. Entertainment and Media

In the entertainment industry, virtual actors or presenters could modify their personas to suit different genres or audience demographics, providing a more tailored entertainment experience.

The ability to dynamically steer the personality of AI models without extensive retraining offers significant advantages in terms of flexibility, cost and user engagement. These potential use cases illustrate how this technology could be leveraged across various industries to enhance the interactivity and personalization of AI-driven applications.

Ethical Considerations

A natural question is whether it is ethical to induce and orthogonalize certain harmful (or even beneficial) personality traits in language models.

Our approach differs from targeting the refusal direction by focusing on personality activation as outlined in [1]. Ethical concerns still arise, particularly around whether it's beneficial to influence language models to exhibit certain personality traits. In this project, we used an uncensored model, but it's possible that targeting specific traits might inadvertently lead to removing refusal capabilities, posing a significant risk.

This approach allows users to elicit toxic or violent responses during runtime. While this capability might be useful in specific scenarios, such as in games where characters need to respond differently based on the situation (as discussed in previous section), it also presents the potential for misuse. This technique could be exploited to produce unintended or harmful responses.

In this project, we identified both "good" and "bad" personality traits. In a production environment, it would be prudent to constrain these personality vectors to prevent misuse and ensure the model's responses remain appropriate and safe.

Another consideration is the broader implications of attributing personalities to language models. Current LLMs interact in human-like ways, which raises questions about the long-term impact of inducing specific personality traits. While this research can provide valuable insights into the capabilities and limitations of LLMs, it also highlights the need for caution. Anthropomorphizing models and encouraging them to exhibit human-like personalities might have unintended negative consequences.

Future Work

This project has merely scratched the surface of the potential applications and methodologies available for manipulating and utilizing LLMs and AI systems. There are several promising directions for future research.

1. Interpretability and Layer Activations
This approach contributes to the interpretability problem by helping understand the inner workings of LLMs. By identifying layer activations associated with specific traits, we gain insights into the black box of LLMs. Future work should focus on deepening this understanding, as the ability to interpret these models is arguably more critical than merely implementing these techniques.

2. Multilingual Models and Cross-Linguistic Personality Traits

Another intriguing research avenue is examining how personality traits manifest in multilingual models. It would be interesting to see if similar personalities in different languages occupy comparable positions in the vector space. Additionally, exploring the discovery of new personality traits in one language that lack a direct counterpart in another could provide fascinating insights into linguistic and cultural differences.

3. Performance Impact of Different Personality Traits

It would be interesting to investigate whether different personality traits affect the overall or specific performance of the model. For example, does a shy personality enhance coding capabilities? Is a mentor-like personality more effective in conversational tasks? Understanding these correlations could optimize LLMs for various applications based on desired traits.

4. Long Context Windows and Conversational Performance

Another area worth exploring is how models with induced or ablated personality traits perform in conversations with long context windows. This could reveal how stable and coherent these personality traits remain over extended interactions and whether they influence the model's ability to maintain context and coherence.

5. Ethical and Safety Considerations

Of course, future work should also address the ethical implications of inducing specific traits in LLMs. Research should aim to develop guidelines and safeguards to prevent misuse and ensure that these technologies are used responsibly.

These are just a few practical approaches that can be pursued in future work and research.

Concluding Remarks

In conclusion, this research demonstrates how *activation engineering* can dynamically adjust personality traits in LLMs. By manipulating activation vectors, we can personalize AI behavior in real time, enhancing applications in various fields such as consumer products, robot companions and personal assistants.

Our findings emphasize both the potential and ethical considerations of this technology. Future research should focus on improving interpretability, exploring different models and ensuring ethical guidelines to prevent misuse.

Acknowledgments

I would like to express my deepest appreciation to the Wolfram Summer School staff for their tireless dedication in making this program a reality. Their commitment has been truly inspiring.

A special thank you goes to my mentor, Vladimir Grankovsky, whose guidance and support were instrumental in bringing this project to fruition. His expertise and patience were invaluable all throughout.

I'm also grateful to the other staff members and other mentors who offered their assistance and insights along the way. Their collective wisdom has significantly enriched my experience.

Finally, I want to convey my heartfelt gratitude to Stephen Wolfram. Our conversations were not only enlightening but also profoundly stimulating. I'm deeply thankful for the time and attention he devoted to my work.

References

1. A. Arditi, et al. (2024), "Refusal in LLMs Is Mediated by a Single Direction," *LessWrong*. www.lesswrong.com/posts/jGuXSZgv6qfdhMCuJ/refusal-in-llms-is-mediated-by-a-single-direction.

2. TurnTrout, et al. (2023), "Steering GPT-2-XL by Adding an Activation Vector," *LessWrong*. www.lesswrong.com/posts/5spBue2z2tw4JuDCx/steering-gpt-2-xl-by-adding-an-activation-vector.

3. Y. Dong, et al. (2023), "SteerLM: Attribute Conditioned SFT as an (User-Steerable) Alternative to RLHF." arxiv.org/pdf/2310.05344.

Access the Full Code

Scan or visit wolfr.am/WSS2024-Allbert.

Cite This Notebook

"Identifying and Manipulating Personality Traits in Large Language Models (LLMs)"
by Rumi Allbert
Wolfram Community, STAFF PICKS, July 9, 2024
community.wolfram.com/groups/-/m/t/3210605

Particles in a Box: Kinetic Gas Theory and Simulations

JAKUB TRZASKA

The aim of this project is to give intuitive insight into the elusive nature of a gas contained in a box through the use of the kinematic approach. We investigate the dynamics of a 1D hard-sphere system, also referred to as a hard-rod gas. By employing the event-driven simulation method, we compute the evolution of a system of colliding particles and visualise the properties of the model. We model a container with two species of gas-particles separated by a wall and explore the dependency on relevant parameters.

1. Hard Spheres

1.1 Introduction and Setup

1.1.1 Details of the Model

The system of interest will be a one-dimensional container filled with particles, exhibiting the behaviour of a gas. To be exact, the particles will be modelled as hard spheres—that is, described by the following interaction potential (with σ being the radius of the particle):

$$u\left(x\right) = \begin{cases} \infty & 0 < x \le \sigma \\ 0 & x > \sigma \end{cases},$$

which conveys the key idea that the particles in this model do not repel or attract each other, but rather collide at direct contact with each other.

Now imagine the same gas-filled container divided in the middle by a movable wall or a piston, reacting to the impacts of particles on both sides. The particles on the left side of the container will be referred to as "species α" or "A," and the particles on the right as "species β" or "B." The two species can have different properties, come into contact with each other and interact directly or be separated by a wall-particle, mediating the interactions. We should keep in mind that in a 1D hard-sphere gas, only adjacent particles can collide and no particles can pass through each other. In other words, the particles are in a fixed order.

1.1.2 Coding the Simulation

The core of our approach at the solution is the event-driven method of simulation. Instead of solving for the positions and velocities of particles at every fixed time step, we are going to compute the time until all possible future collisions for given initial conditions and then evolve the system analytically up to that point. Notably, this means that each time step of an event-driven simulation is of a different length. This realisation of the simulation allows for faster computation and, in some cases, avoidance of numerical errors.

Here we include relevant secondary functions, necessary to successfully compute a time step of a simulation.

Time until a collision between two given particles is handled by the following function:

```
In[•]:= CollisionTime[x1_, x2_, v1_, v2_, r1_, r2_] := Module[{Δx, Δv, Δt, d},
          d = r1 + r2;
          Δv = v2 - v1;
          Δx = x2 - x1;
          Δt = If[ Δv Δx ≥ 0, Infinity, -(Δv Δx + Sqrt[Δv² d²]) / Δv² ]
       ];
```

Update of the position of a particle with a constant velocity:

In[ ]:= **Move[x_, v_, t_] := Module[{}, x + t v];**

To realise a collision, we need a function which will compute the final velocities of colliding particles. We implement it as follows:

In[ ]:= **doCollision[x1_, x2_, v1_, v2_, r1_, r2_, m1_, m2_] := Module[{Δx, Δv, v1New, v2New},**
$$\Delta v = v2 - v1;$$
$$\Delta x = x2 - x1;$$
$$v1New = v1 + \frac{2\,m2}{m1 + m2}\, \Delta v;$$
$$v2New = v2 - \frac{2\,m1}{m1 + m2}\, \Delta v;$$
$$\{v1New, v2New\}$$
];

Following, we introduce the master function for the simulation of a hard-sphere gas. It allows us to model two different particle species separated by a wall-particle. It also contains the implementation of a spring-wall, which will be discussed later (see code appendix for full details):

In[ ]:= **HSsim[]**

The code for this function (and other functions, which take up a lot of space) is included in the code appendix at the end of the post.

1.2 Numerical Experiments

1.2.1 Setup

Having constructed the hard-sphere model framework, we can finally run proper simulations. While we might start with an example with particles of one species bouncing around in a box, the goal is to simulate two species separated by a wall.

For the case with a wall-particle, there are two approaches that come to mind: a lightweight wall that is going to act as a probe at the interface between two gases, or a heavy wall that will possess its own significant momentum and will influence the dynamics of the system.

In the following sections, we present examples of system evolution for these two approaches.

For all cases in this section, unless specified otherwise, the simulation parameters are as follows (these will be referred to as *standard parameters*):

In[ ]:= **nA = 1;**
nW = 0;
nB = 1;

RA = 5;
RW = 20;

```
        RB = 5;

        width = 100;
        vmax = 10;

        mA = 1;
        mW = 10^3;
        mB = 1;

        wallSpring = False;
        springConst = 1;

        steps = 10;
```

Now we can construct an input list, which will then be passed to the master function:

```
In[·]:=  input := <|
            "PosAlpha" → Table[{ 2 RA + 3 RA (i − 1)}, {i, 1, nA}],
            "PosWall" → Table[{ width / 2 + 2 RW (i − 1)}, {i, 1, nW}],
            "PosBeta" → Reverse[Table[{ width − 2 RB − 3 RB (i − 1)}, {i, 1, nB}]],

            "VelAlpha" → Table[ {vmax} , {i, 1, nA}],
            "VelWall" → Table[{0}, nW],
            "VelBeta" → Table[ {vmax}, {i, 1, nB}],

            "AlphaRadius" → RA,
            "WallRadius" → RW,
            "BetaRadius" → RB,

            "MassAlpha" → mA,
            "MassWall" → mW,
            "MassBeta" → mB,

            "Steps" → steps,
            "BoxWidth" → width,

            "SpringConst" → springConst,
            "WallSpring" → wallSpring
          |>
```

The function used to plot the trajectories of particles is defined as follows (see code appendix for full details):

```
In[·]:=  PlotEvolution[]
```

Now we can finally simulate the evolution of a simple, two-particle system:

```
In[·]:=  simulationResult = HSsim[input];
```

Plot the result of the simulation. The particle trajectories are colored randomly, to be more easily distinguishable:

In[ ]:= **PlotEvolution[simulationResult]**

Out[ ]=

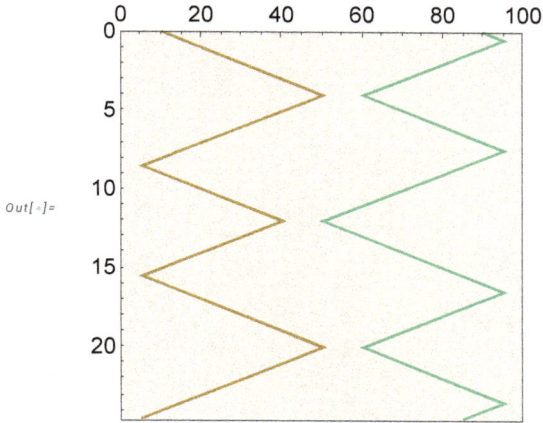

In the previous plot, the initial state is at the top and the system evolves in time along the vertical axis. We can also consider a system with one *heavy wall* (portrayed by the gray line) and a few more particles on each side, while also making the box a bit wider (note the greater radius of the wall-particle):

In[ ]:= **nA = 3;**
nW = 1;
nB = 3;

width = 200;
steps = 18;
input;

PlotEvolution[HSsim[input]]

Out[ ]=

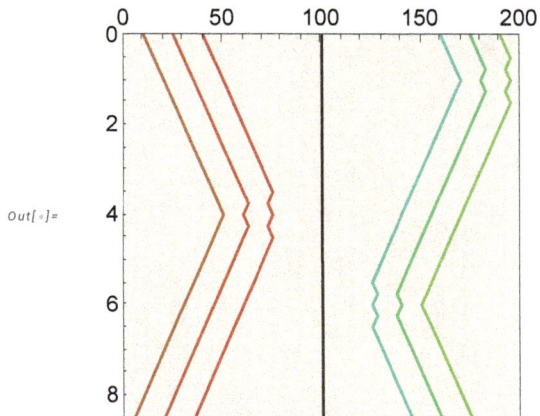

We can now expand our simulation to a greater number of particles. The purpose of giving the particles the same initial velocities is to eventually observe how they evolve into disorder:

```
In[ ]:= nA = 10;
        nW = 1;
        nB = 10;

        width = 400;
        steps = 500;
        input;

        PlotEvolution[HSsim[input]]
```

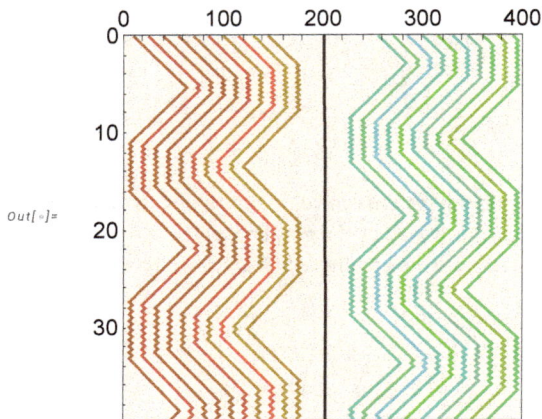

Clearly, the core of our interest will be much longer simulations, where we can see both the desired disorder and the interesting behaviour of the wall-particle:

```
In[ ]:= nA = 10;
        nW = 1;
        nB = 10;

        mW = 10^3;

        width = 500;
        steps = 4000;
        wallSpring = False;
        input;

        longSimResult = HSsim[input];

        PlotEvolution[longSimResult]
```

$Out[\circ]=$

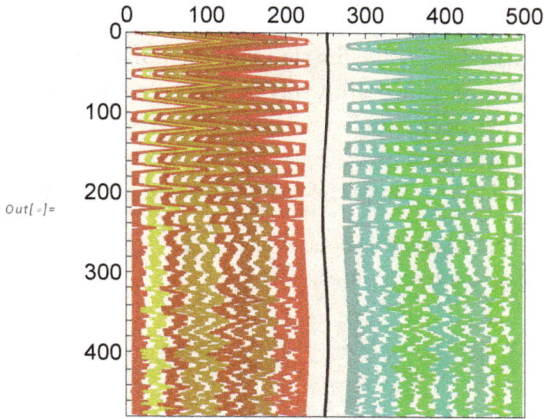

1.2.2 Energy Conservation Plots

One of the collective properties of the system, which we can easily extract from the simulation, is the *kinetic energy* of each of the components. After all, we track the evolution of velocities of every particle in the simulation. By combining this data, we will be able to see how the energy is transferred during collisions, but also check if the total energy is conserved (which it ought to be, if the simulation works as expected).

Let us plot the evolution of the system, using the previous simulation as input (see code appendix for full details):

$In[\circ]:=$ **PlotEnergy[]**

$In[\circ]:=$ **PlotEnergy[longSimResult]**

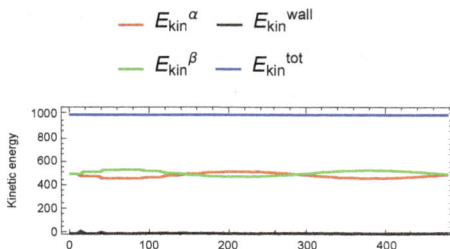

$Out[\circ]=$

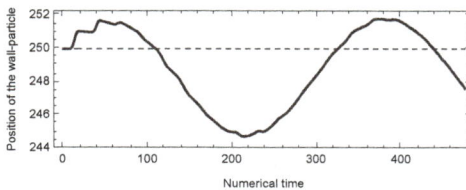

The motion of the wall is periodic, which has been cross-checked in numerous simulations spanning across a wide range of parameters (the results of longer simulations weigh much more). We will make use of this fact in the following sections.

1.2.3 Varying the Radius

Several simulations of 20,000 steps were carried out using the standard set of masses (width = 500, $n_\alpha = n_\beta = 10$ and 1 wall-particle—only the radius was changed between subsequent runs). For each of them, a sine wave has been fitted to the trajectory of the wall-particle to determine the period of wall oscillations. Surprisingly, we found the relation between the period of motion of the wall and the radius of the gas-particles to be linear!

The results of the simulations are given here, but an ambitious reader is more than encouraged to run the simulations again and compare the outcomes.

Final fit of the model to the simulation data:

```
In[ ]:= testRadii = {1, 3, 5, 7};
periodFitResult = {51.83, 43.94, 33.69, 23.17};
data = Transpose[{testRadii, periodFitResult}];

plotOne = ListPlot[data, PlotLegends → Placed[{"Simulation data"}, {0.8, 0.8}],
        PlotStyle → {Red}, PlotMarkers → {Automatic, Medium}];
model = a x + b;
fit = FindFit[Transpose[{{1, 3, 5, 7}, {51.83, 43.94, 33.69, 23.17}}], model, {a, b}, x]
FindRoot[model /. fit, {x, 5}]

plotTwo = Plot[model /. fit, {x, 0.5, 8},
        PlotLegends → Placed[{"Linear model fit"}, {0.8, 0.8}], PlotStyle → Black];
plotThree = Plot[model /. fit, {x, 0.5, 13}, PlotStyle → {Black, Dashed}];

Show[{plotTwo, plotThree, plotOne}, Frame → True,
    FrameLabel → {"Radius of species α and β particles", "Period of wall oscillations"},
    PlotRange → {{-1.2, 13}, {-5, 55}}, ImagePadding → Automatic,
    ImageSize → Medium, AspectRatio → 3/4, AxesOrigin → {0, 0}]
```

Out[ ]= {a → −4.8115, b → 57.4035}

Out[ ]= {x → 11.9305}

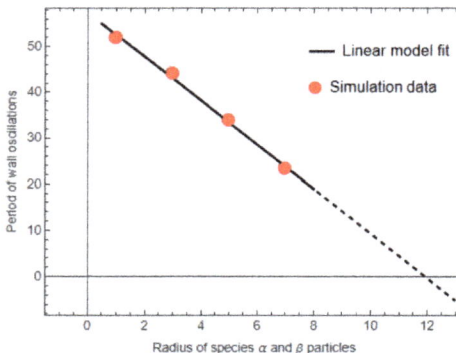

It should be noted that the zero of the fitted function roughly corresponds to a radius of a gas-particle which (for the total number of gas-particles, 10) makes the particles fill all of the available space in the box.

1.2.4 The Appearance of Power Law

A similar scan has been performed with 20,000 step simulations (width = 500, $n_\alpha = n_\beta = 10$ and standard radius, this time changing only the mass of gas-particles surrounding the wall-particle). By plotting the results, we quickly see a power law–like behaviour. And indeed, the slope of the line in a log-log plot has been determined to be –1/2, which means:

$\tau_{\text{wall}} \propto m^{-1/2}$,

$m = m_\alpha = m_\beta$,

... where τ_{wall} is the period of wall oscillations. Here we present the final fit of the model relating the period of wall oscillations and gas-particle mass.

Final fit of the model to the simulation data:

```
In[ ]:=  testMasses = {1, 2, 5, 10};
         periodFitResult = {33.69, 23.52, 15.22, 10.80};
         data = Transpose[{testMasses, periodFitResult}];

         plotOne = ListLogLogPlot[data, PlotLegends → Placed[{"Simulation data"}, {0.8, 0.8}],
             PlotStyle → {Darker[Green]}, PlotMarkers → {Automatic, Medium}];

         model = a x^{-1/2};
         fit = FindFit[data, model, {a}, x]
         plotTwo = LogLogPlot[model /. fit, {x, 0.8, 15}, PlotStyle → {Black},
             PlotLegends → Placed[{"—a/√x— model fit"}, {0.8, 0.8}]];

         Show[{plotTwo, plotOne}, Frame → True,
             FrameLabel → {"Mass of species α and β particles", "Period of wall oscillations"},
             PlotRange → {{-0.5, Log[15]}, {2, Log[40]}}, ImagePadding → Automatic,
             AspectRatio → 3/4, Axes → False, ImageSize → Medium]
```

```
Out[ ]=  {a → 33.635}
```

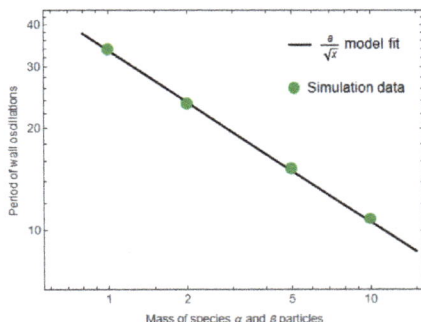

Although interesting, the theoretical foundations of this relation, and the one found in the previous section, will not be explored further in this simulation-oriented post.

1.2.5 Asymmetric Case

Up to this point, we considered particles of the same kind on both sides of the wall. The framework allows us, however, to choose the parameters of both gases separately. In the following, we consider an asymmetric case with a lightweight wall-particle, acting as a probe between two gas species.

Here, we give an example of a setup of an asymmetric simulation:

```
In[ ]:=  nA = 20;
         nW = 1;
         nB = 10;

         RA = 5;
         RW = 20;
         RB = 10;

         mA = 1;
         mW = 0.1;
         mB = 1;

         width = 800;
         steps = 3000;
         wallSpring = False;
         input;

         lightWallSimResult = HSsim[input];

         PlotEvolution[lightWallSimResult]
```

Out[ ]=

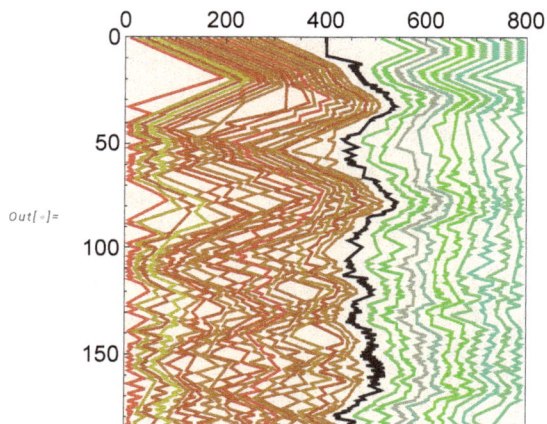

We can also take a look at the energy and wall position plots:

In[ ]:= **PlotEnergy[lightWallSimResult]**

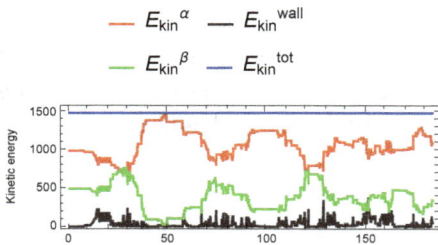

Out[ ]=

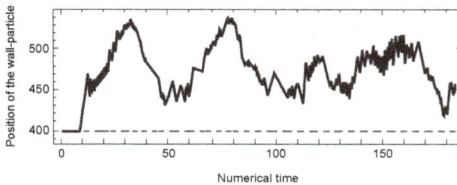

Due to the use of the lightweight wall approach, the behaviour of the wall-particle is much more erratic; however, periodic motion can still be observed. In principle, tracking its position allows us to see the asymmetric system reaching an equilibrium.

1.3 Extracting the Equation of State

Let us take a detour and search for a physical meaning of the hard-sphere gas model. For the sake of simplicity, in the following sections we assume $k_B = R = 1$, where k_B is the Boltzmann constant and R is the gas constant.

1.3.1 Virial Expansion

The basic idea behind the virial expansion is to expand the equation of state. To avoid going into detail unnecessary for this post, we can write said expansion as follows:

$$\frac{p}{\rho} = T \left(1 + \rho B_2 \left(T\right) + \rho^2 B_3 \left(T\right) + ...\right),$$

where p is pressure, T is temperature and $\rho = \frac{N}{V}$ is the number density. We notice right away that the first term of the expansion replicates the ideal gas law:

$$p = T\rho \qquad \Longrightarrow \qquad pV = N\,T.$$

Subsequent terms allow us to express the equation of state of a given system more accurately, if only we can compute the relevant virial coefficients. For instance, the general expression for the second coefficient in the 3D case reads:

$$B_2 \left(T\right) = -\frac{1}{2} \left(4\,\pi\right) \int r^2 \; \mathrm{d}\,r \; f\left(r\right),$$

where $f(r)$ is the Mayer's function:

$$f(r) = \exp\left(-\frac{u(r)}{T}\right) - 1,$$

$u(r)$ being the interaction potential. Naturally, the higher coefficients require us to take many-body interactions into account, making the computations more complicated. Luckily for us, in the case of the 1D hard-sphere potential (also called hard-rod), these coefficients can be easily obtained explicitly, giving the following:

$$B_2 = \sigma,$$
$$B_3 = \sigma^2 = B_2{}^2,$$

where σ is the radius of the particle. It should be noted that virial coefficients are, in general, functions of temperature. Using the above, we obtain the following equation of state (EoS):

$$p = T\left(\rho + \sigma\rho^2 + \sigma^2\rho^3\right).$$

That is simple and elegant enough. Now let us see if we can somehow apply this to the hard-sphere system simulation.

1.3.2 Comparing Theory and Simulation

Naturally, we would like to see our simulations being confronted with appropriate theory to assess their validity and applicability. In a sufficiently long simulation, we expect the system to reach equilibrium. In that case, the pressure on both sides of the wall should be equal, thus:

$$T_\alpha(\rho_\alpha + \sigma_\alpha\,\rho_\alpha{}^2 + \sigma_\alpha{}^2\,\rho_\alpha{}^3) = T_\beta\left(\rho_\beta + \sigma_\beta\,\rho_\beta{}^2 + \sigma_\beta{}^2\,\rho_\beta{}^3\right).$$

Additionally, we will assume the temperatures at equilibrium to be equal. We would then expect the following ratio to be close to unity:

$$\frac{\left(\rho_\alpha + \sigma_\alpha\,\rho_\alpha{}^2 + \sigma_\alpha{}^2\,\rho_\alpha{}^3\right)}{\left(\rho_\beta + \sigma_\beta\,\rho_\beta{}^2 + \sigma_\beta{}^2\,\rho_\beta{}^3\right)} \approx 1.$$

To determine the exact range of applicability of this approach, we need to perform an appropriate scan of the simulation parameter space. This is requires a large simulation base, and is an ongoing part of this project.

1.4 Adding a Spring to the Wall

This section is devoted to modelling the wall-particle as a harmonic oscillator. Initially, the particle will reside in the minimum of a harmonic potential, and every collision with adjacent gas-particles will cause it to oscillate around the position of equilibrium.

1.4.1 Setting Up the Simulation

Part of the code has already been implemented in the master function introduced at the beginning, but there are several crucial functions left to be defined.

Time until a collision between a gas-particle and a spring-wall-particle requires a slightly more involved function:

```
In[•]:= CollisionTimeSpringWall[x0wall_, x0_, v0wall_, v0_, Rwall_, R_, k_, M_, ξ_] :=
        Module[{Δt, sol},
            (* ξ – equilibrium position of the spring (initial positions of the wall) *)
            (* Find the solution for given initial conditions : *)
            sol = DSolve[{x''[t] == -k/M (x[t] - ξ), x[0] == x0wall, x'[0] == v0wall}, x[t], t];
            Δt = NSolve[{Sign[x0 - x0wall] (-(x[t] /. sol[[1]]) + v0 t + x0) == Rwall + R, t > 0}, t];
            If[Δt == {}, Infinity, t /. Δt[[1]]]
        ];
```

To update the position of the wall, we need a solution for the harmonic oscillator. Just for the purpose of visualisation, we may compute additional positions along the way (useful especially with few particles, and hence few collision time steps). The function handling the update should also return the final velocity of the wall-particle. All that results in the following function:

```
In[•]:= MoveSpringWall[x0wall_, v0wall_, k_, m_, ξ_, tFinal_] :=
        Module[{sol, PosTimeList, div},
            (* ξ – equilibrium position of the spring (initial positions of the wall) *)
            (* Number of divisions of given timestep *)
            div = 50;

            (* Find the solution for given initial conditions *)
            sol = DSolve[{x''[t] == -k/m (x[t] - ξ), x[0] == x0wall, x'[0] == v0wall}, x[t], t];
            (* Divide the solution into smaller timesteps *)
            PosTimeList =
                Table[{N[(x[t] /. sol[[1]]) /. t → (i tFinal)/div], tFinal/div}, {i, 1, div}];

            (* {final position, all positions during motion, final velocity} *)
            {PosTimeList[[-1]][[1]], PosTimeList, D[x[t] /. sol[[1]], t] /. t → tFinal}
        ];
```

We also need to make a few minor changes to the plotting function to account for the additional time steps of the oscillating wall-particle:

```
In[•]:= PlotSpringWall[]
```

1.4.2 Example Results

Let us run a simple simulation to test the newly introduced spring-wall functionality:

```
In[•]:= nA = 1;
        nW = 1;
        nB = 0;
```

```
mW = 10^1;

vmax = 100;
width = 300;
steps = 20;
wallSpring = True;
springConst = 10^2;

input;
simResult = HSsim[input];
```

... and plot the result:

In[ ]:= **PlotSpringWall[simResult]**

Out[ ]=

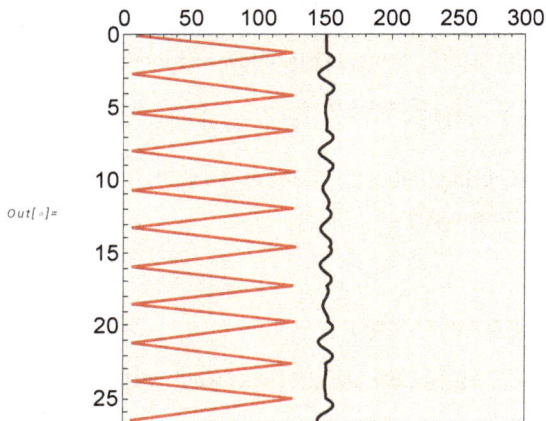

We also introduce the energy-plotting function, with minor changes compared to the previous one:

In[ ]:= **PlotEnergySpringWall[]**

Now we can plot the kinetic energies of the components of the system. The sharp changes in the kinetic energy of the oscillating wall are not physical, and are a result of a lesser number of global time steps of the simulation:

In[ ]:= **PlotEnergySpringWall[simResult]**

Out[ ]=

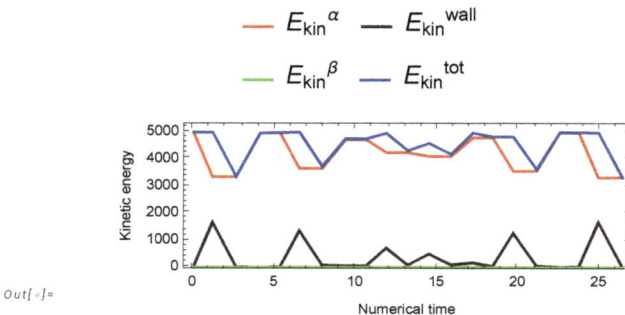

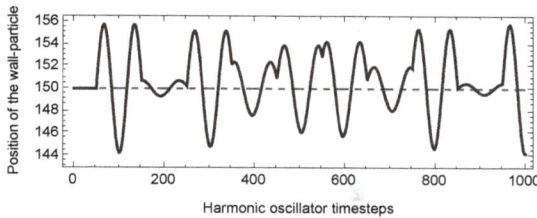

Harmonic oscillator timesteps

As we should expect, total kinetic energy is not conserved—it is partially stored in the deformed spring, in the form of potential energy. We can clearly see the regions of harmonic motion and the points of collision with a gas-particle.

We are in the process of analysing this simulation functionality further. Many other numerical experiments can be performed, which a curious reader is also encouraged to try.

Concluding Remarks

As we have shown, even the simplest one-dimensional model of a gas has a lot to offer in terms of rising complexity, programming challenges and physics that we can extract from it.

In terms of outlook for the future, there is certainly a lot more to experiment on. We can, for instance, model a solid wall as a lattice of masses connected by springs, code for which is already partially done. We can then place such a wall in a container, separating two gases, similar to what has been shown in this post. An extension of the hard-sphere event-driven simulation to the 2D case is also possible, but due to additional complexity (both model- and code-wise) it deserves its own separate project.

Acknowledgments

I want to express my gratitude to my team from the International Physicists' Tournament—I wouldn't be a part of the Wolfram Summer School without them. I would also like to thank the dean of student affairs at my faculty, Krzysztof Turzyński, who made my participation in the Summer School possible. Lastly, I want to acknowledge Stephen Wolfram for suggesting the idea for this interesting project.

Reference

1. Stephen Wolfram (2023), "Computational Foundations for the Second Law of Thermodynamics," *Stephen Wolfram Writings*. writings.stephenwolfram.com/2023/02 /computational-foundations-for-the-second-law-of-thermodynamics.

Access the Full Code

Scan or visit wolfr.am/WSS2024-Trzaska.

Cite This Notebook

"Particles in a Box: Kinetic Gas Theory and Simulations"
by Jakub Trzaska
Wolfram Community, STAFF PICKS, July 9, 2024
community.wolfram.com/groups/-/m/t/3210337

Search for Rules that Preserve Geometric Structure in Wolfram Models

FERGUS NOBLE

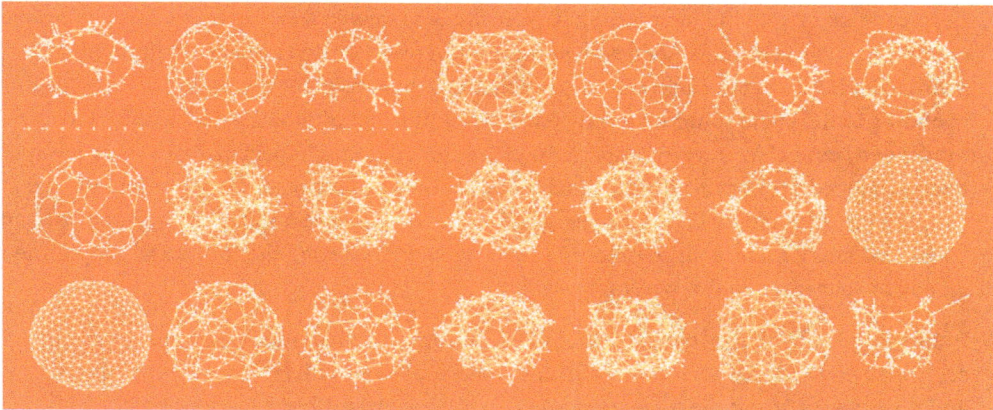

The Wolfram model is a class of models consisting of a graph (or hypergraph) and a rewriting rule. It has been shown by [1] and [2] that the Wolfram model can exhibit behavior corresponding to general relativity in the continuum limit; however, it remains unknown precisely which rules show this correspondence. In this project, I conducted a systematic search over the space of possible rewriting rules to identify rules that preserve the geometric structure of graphs that correspond to static vacuum solutions of the Einstein equations, a necessary condition for these rules to satisfy general relativity. I considered rules involving two and three edges on simple directed graphs corresponding to flat Minkowski and Schwarzschild solutions. From a search space of 49,116 rules, I identified 32 that preserve geometric structure and may be candidates for modeling general relativistic behavior.

Introduction

A foundational principle of the Wolfram Physics Project is that spacetime is discretized at small scales and may be represented as a (hyper)graph structure that evolves according to the action of a hypergraph rewriting rule [1]. This concept is known as the Wolfram model. This contrasts with the traditional view in general relativity, where spacetime is represented as a continuous Riemannian manifold evolving according to a set of partial differential equations (the Einstein equations). The question then arises of how these views can be reconciled. It has been shown that a latent manifold structure can be recovered from certain hypergraphs in an appropriate continuum limit [1, 2], and that certain classes of rewriting rules result in hypergraph evolution that satisfies the Einstein equations. However, not all rewriting rules will produce hypergraphs with a manifold structure. It remains an open problem in the Wolfram Physics Project to understand which rules give rise to manifold structure and satisfy the Einstein equations in the appropriate limit.

Consider a manifold that corresponds to a static solution of the Einstein equations. One necessary (but not sufficient) property for a suitable rewriting rule is that its action on a hypergraph with such a manifold structure should preserve the overall manifold structure upon repeated application. In this project, we conduct a systematic search for rewriting rules that preserve static manifold structure for flat Minkowski spacetime as well as for the Schwarzschild solution.

In a Wolfram model, the hypergraph is typically constructed starting with a minimal initial condition, such as a single node with several self-loops. For this project, I will instead prepare an initial hypergraph that is constructed to have a latent manifold structure corresponding to the static solution of interest. These initial conditions may be considered a model of a "patch" (connected subgraph) taken from a larger hypergraph that may have been produced through rewriting. Since we are only considering static solutions, we can simplify the problem by taking a hypergraph corresponding to a spacelike hypersurface and considering its evolution under the rule rather than considering the full causal graph (the analog of spacetime in the Wolfram model).

In this project, I will first consider the role of the graph "microstructure" and establish criteria to bound the search space of rules. Specifically, we will look at rules involving two and three edges on simple directed graphs. I will then define heuristics to test rules for physical behavior. Subsequently, I will conduct an exhaustive search over the space of rules, applying the rules to four example initial graphs corresponding to flat space in two dimensions and evaluating the heuristics for each rule. Rules that pass the heuristic tests are then applied to a larger input graph corresponding to a Schwarzschild solution and reevaluated. Additional tests are performed on the rules to check for triviality (isomorphism) and preservation of the graph's Wolfram–Hausdorff dimension [1]. From these tests, 32 rules are identified from the initial search space of 49,116 rules.

Microstructure and Macrostructure

We have discussed that a hypergraph may have a corresponding latent manifold structure that we can consider the *macrostructure* of the hypergraph; however, there are many possible hypergraphs that have equivalent coarse-grained macrostructure. These graphs are differentiated in terms of their microstructure, i.e. the detailed graph structure at short range. For example, consider the following graphs:

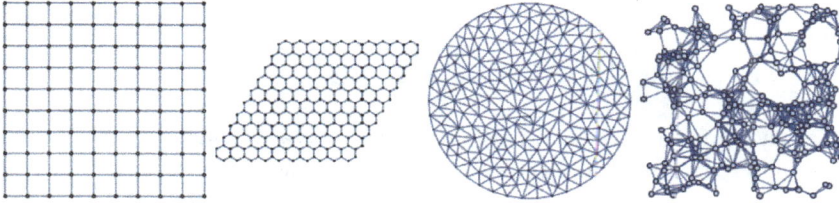

Each of these graphs has a macrostructure corresponding to two-dimensional Euclidean space, but they differ in their microstructure. The first has a rectangular grid structure, the second is a hexagonal grid, the third is a triangular mesh and the fourth is produced by connecting a set of random vertices ("random sprinkling").

Hypergraphs generated from rewriting rules from a minimal initial condition often show emergent microstructure [1, ch. 4.1]. For example, the rule:

$$\{\{x, x, y\}, \{x, z, u\}\} \rightarrow \{\{u, u, z\}, \{y, v, z\}, \{y, v, z\}\}$$

… produces the following hypergraph after one thousand steps:

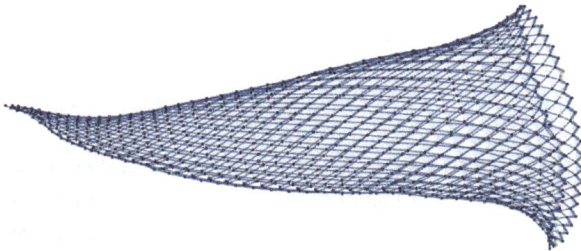

A clear repeated microstructure has become apparent.

It is possible that a rule may be manifold-preserving given one particular microstructure but not another. As we are preparing model initial conditions that were not generated via rewriting, it is possible that the arbitrary choice of microstructure may impact our conclusions about the suitability of a rule. In this project, we will also examine how the choice of microstructure impacts the behavior of various rules.

The other major way in which our model differs from a hypergraph produced through rewriting is at the boundary. It is possible that the action of the rules at the boundary will differ from the behavior of the subgraph within a larger graph.

Enumerating the Space of Rules

There are infinitely many possible rewriting rules. To perform a systematic search, we must constrain the search space. In this project, I will apply the following constraints:

- All of our methods for preparing initial condition hypergraphs will generate simple directed graphs. Therefore, there is no need to include rules involving hyperedges of more than two vertices, i.e. we can restrict our search to binary rules with signature [1. ch 3.1] of the form $N_2 \to M_2$.

- Rules that introduce new vertices or edges will quickly lead to exponential growth. Therefore, we can further restrict our search to rules with the signature $N_2 \to N_2$.

- It has been hypothesized that increasing the complexity of the rule (i.e. higher values of N) will not increase the complexity of the behavior above some low threshold [1, ch 3.11]. Previous results have shown $2_2 \to 5_2$ does not show significantly more complex behavior than $2_2 \to 4_2$, so we will not consider N > 4.

There are 562 rules with signature $2_2 \to 2_2$, and 48,554 rules with signature $3_2 \to 3_2$. Using the formula from [1, ch 3.2] for an estimate of the number of rules using Bell's number, we can estimate the number of rules with signature $4_2 \to 4_2$ is of order 10^7. As this is too large of a space to search systematically, we will limit ourselves in this project to rules with signatures $2_2 \to 2_2$ and $3_2 \to 3_2$.

Criteria for Rule Selection

Testing the resultant hypergraph after rule application to determine if the manifold macrostructure has been preserved is a nontrivial task. There are techniques to calculate the curvature of a hypergraph [1, 2], but they can be computationally expensive and may require large graphs to converge sufficiently to the continuum limit. It is also expected that a large fraction of possible rules will result in more obvious pathologies. To provide a fast and computationally efficient way to down-select the set of rules, we can use a set of heuristic tests to determine if a rule exhibits nonphysical behavior after only a few rounds ("generations") of rule application.

The rule should:

1. Not result in disconnected subgraphs.

2. Not cause the number of vertices and/or edges to grow exponentially.
   - Note: It is possible to imagine a scheme where the graph "subdivides" while preserving macroscopic structure, but this would not be compatible with the way typical Wolfram models equate graph distances with metric distances, and would require a more complex metric on the hypergraph to preserve lengths and volumes over time. I will exclude this possibility for now.

3. Result in a nonzero number of updating events.

- Note: It is possible that a suitable rule would behave such that it results in no updating events in "empty" space but does behave correctly in the presence of "particles" or other dynamical structures. As in this project we only model static vacuum solutions, it is not possible to further evaluate these rules for suitability.

4. Preserve approximate dimensionality.

- We will measure this by computing the graph diameter and constraining the graph diameter not to shrink below a threshold compared to its initial value. As the number of vertices is approximately constant, this crudely approximates that dimensionality is preserved.

For this project, I will use a threshold for graph diameter of 50% of the initial value and a threshold for vertex and edge growth of 20% compared to the previous generation.

Evolution of an Example Graph

We begin by examining a few concrete examples of rule behavior when applied to the mesh graph:

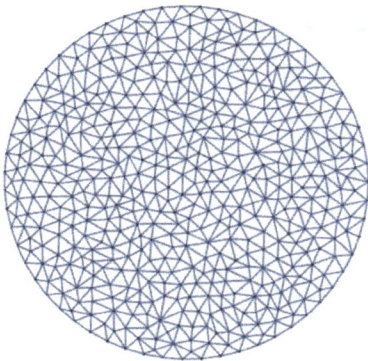

Starting with the rule $\{\{x, y\}, \{x, z\}\} \to \{\{x, z\}, \{y, z\}\}$:

```
In[ ]:= (ruleEx2 = {{x, y}, {x, z}} → {{x, z}, {y, z}}) //
            ResourceFunction["CanonicalWolframModelRule"]) //
        ResourceFunction["WolframModel"] // RulePlot
```

We can tabulate the evolution of the graph for each generation. For each generation, I compute a variety of measures on the graph and check for the previous criteria. If the graph fails one of the criteria, we terminate the evolution. For this rule, the graph passes all tests for the first five generations:

In[ ]:= **RunExperiment[None, graphToWolframModel[graphMesh2D], ruleEx2, 6]**

Out[ ]=

+	Rule	Rule Signature	+	Generation	Graph	Vertex Count	Edge Count	Mean
•		$2_2 \to 2_2$	•	0		257	726	5.649
•		$2_2 \to 2_2$	•	1		257	726	5.649
•		$2_2 \to 2_2$	•	2		257	726	5.649
•		$2_2 \to 2_2$	•	3		257	726	5.649
•		$2_2 \to 2_2$	•	4		257	726	5.649
•		$2_2 \to 2_2$	•	5		257	726	5.649

For the rule $\{\{x, y\}, \{x, z\}\} \to \{\{x, z\}, \{x, w\}, \{y, w\}, \{z, w\}\}$:

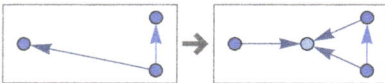

We see that after one generation, the rule fails the Δvertex and Δedge tests as shown in the column "Termination Criteria." This is expected, as the rule introduces a new vertex on each event and therefore will grow exponentially:

In[ ]:= **RunExperiment[None, graphToWolframModel[graphMesh2D], ruleEx1, 5]**

Out[ ]=

+	Rule	Rule Signature	+	Generation	Graph	Vertex Count	Edge Count	Mean
•		$2_2 \to 4_2$	•	0		257	726	5.649
•		$2_2 \to 4_2$	•	1		560	1332	4.757

For the rule $\{\{x, x\}, \{x, x\}\} \rightarrow \{\{x, x\}, \{x, x\}\}$:

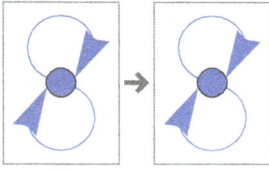

As our input graph does not contain self-loops, this rule will not be applied and there will be no events:

In[ ]:= **RunExperiment[None, graphToWolframModel[graphMesh2D], ruleEx3, 6]**

+	Rule	Rule Signature	+	Generation	Graph	Vertex Count	Edge Count	Mean
•		$2_2 \rightarrow 2_2$	•	0		257	726	5.649
•		$2_2 \rightarrow 2_2$	•	1		257	726	5.649

Out[ ]=

Examples of Graph Evolution

To demonstrate the variety of behavior exhibited by different rules, we can plot some examples of a variety of rules with signature $3_2 \rightarrow 3_2$. Each graph corresponds to the state after the fourth generation:

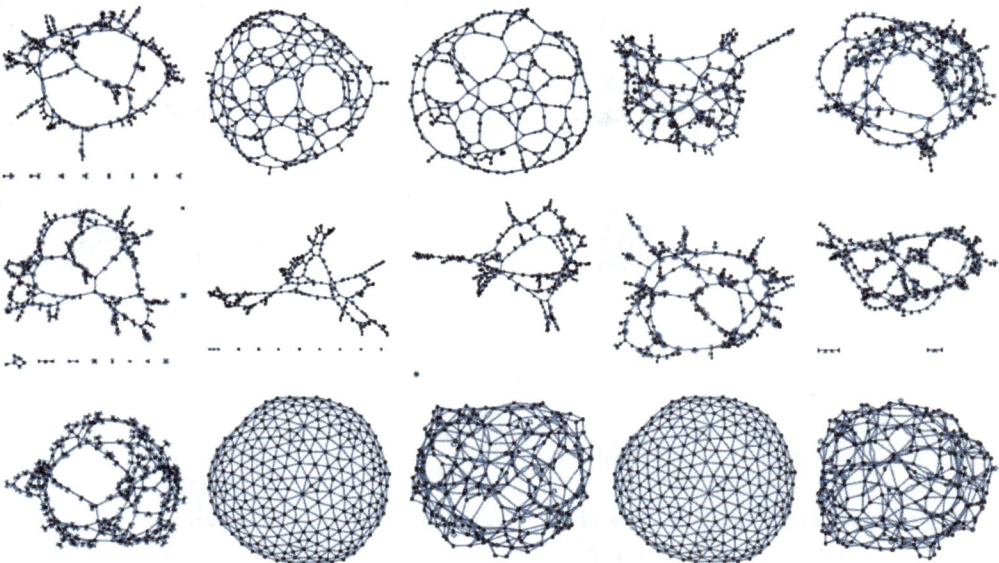

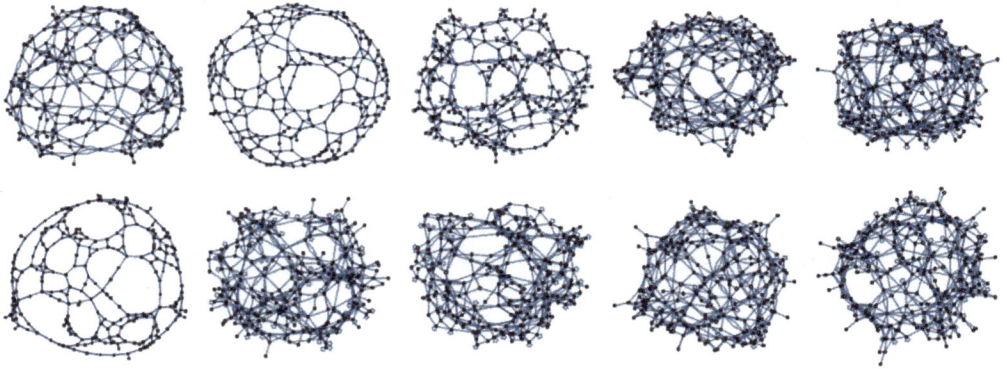

In this figure, we can see a variety of behaviors. Some graphs appear unchanged by the rule. These correspond to rules that transform the graph in some trivial way, such as flipping the direction of the edges. Some rules result in disconnected subgraphs; these will not be carried forward to subsequent generations. Others result in a rich variety of resultant structure.

To see the impact of different choices of microstructure, we can perform the same procedure on the hexagonal grid. This graph shares the same flat macrostructure but shows significantly different behavior:

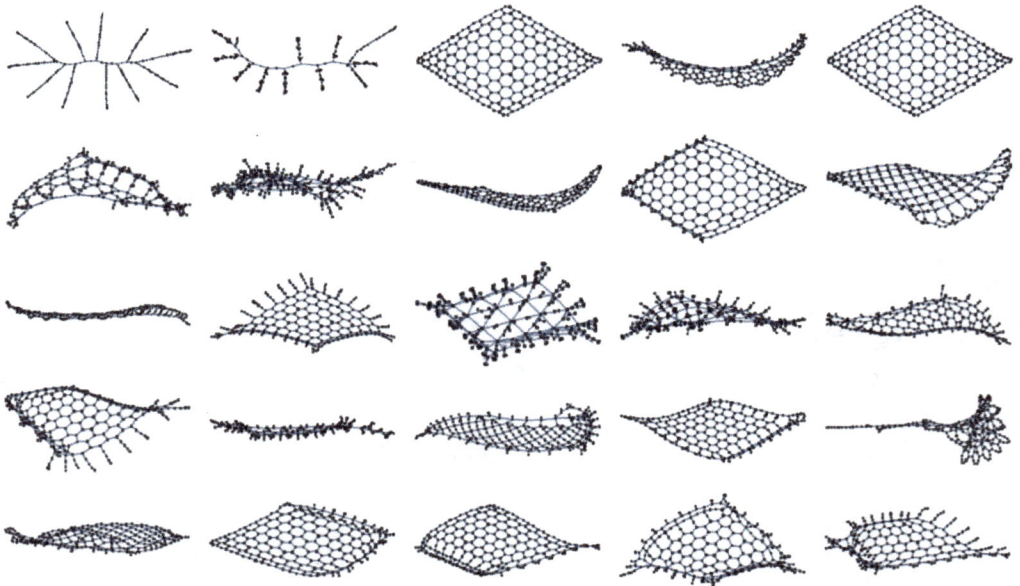

Systematic Search with Flat Two-Dimensional Graphs

For rules with signatures $2_2 \rightarrow 2_2$ and $3_2 \rightarrow 3_2$, the number of rules is few enough that we can perform an exhaustive search on small input graphs. To explore the effect of microstructure, we will start with the four different graphs representing flat space in two spatial dimensions introduced earlier.

The following plot shows the full space of rules with signatures $3_2 \to 3_2$ (48,554 rules) and $2_2 \to 2_2$ (562 rules). In the plots, each square corresponds to a single choice of rule. The rule is repeatedly applied to the input graph. After each generation, the mesh is tested according to the heuristic criteria (see appendix B for details). Rules that survived four generations are deemed to have "passed" and are colored in blue. Rules that were terminated due to violating one of the criteria are colored according to which criterion was violated. Also note the squares in gray, which correspond to rules that did not get applied, i.e. there were no associated events. These represent static conditions on these graphs, but may show dynamic behavior under other circumstances:

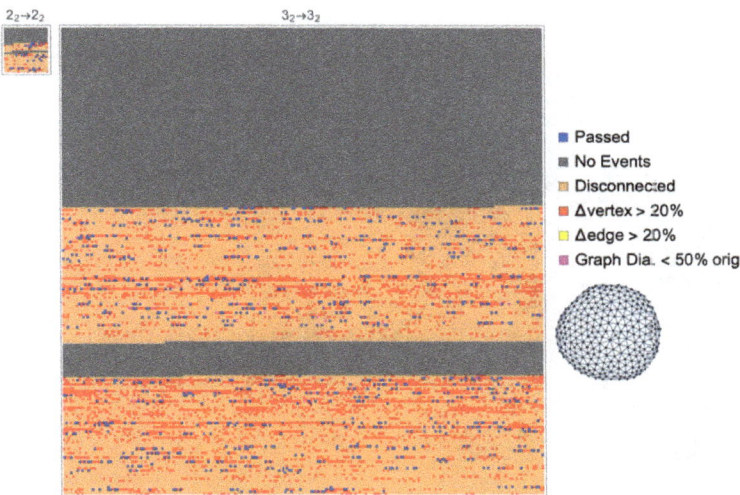

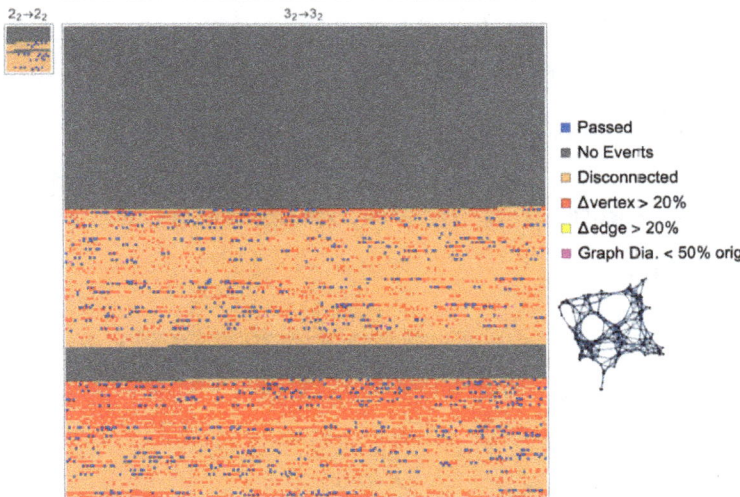

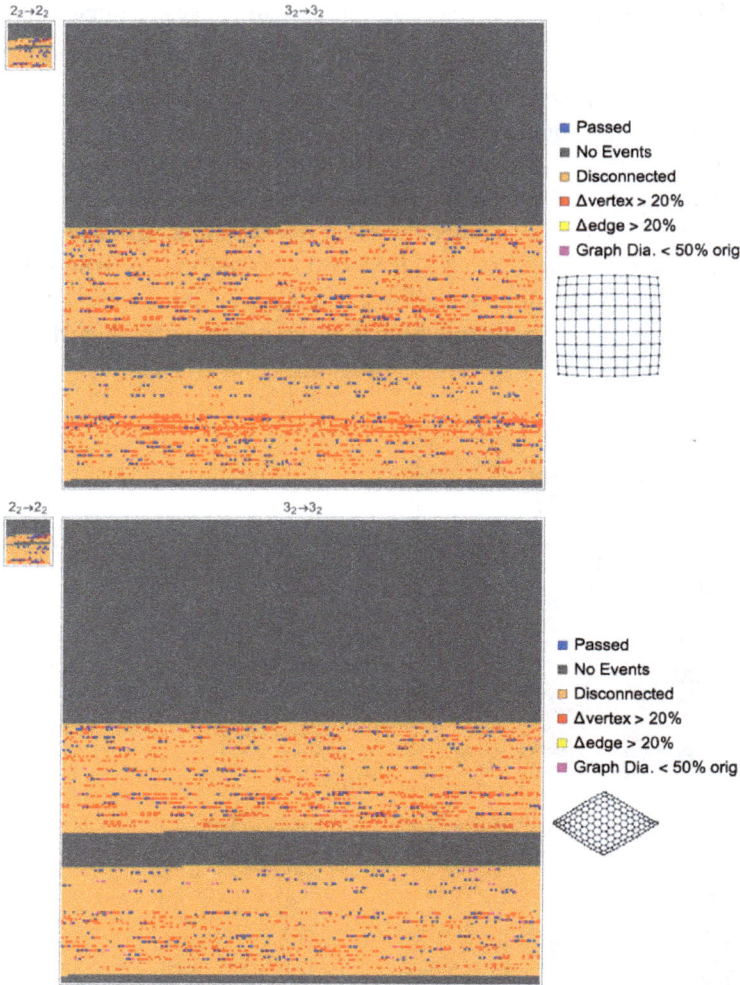

For the first example of the mesh graph, we can compute the percentages of each failure criteria, per signature:

In[ ]:= **TermStatsTable[ds[Select[#["Input Graph Hash"] == 2 784 823 366 451 570 929 &]]]**

Out[ ]=

{{3, 2}} → {{3, 2}}	Disconnected	42.82%
	Graph Dia. < 50% orig	0.06385%
	No Events	45.74%
	Δvertex > 20%	44.77%
	None	2.053%
{{2, 2}} → {{2, 2}}	Disconnected	46.98%
	No Events	39.32%
	Δvertex > 20%	43.77%
	None	5.338%

We can immediately observe some features of the distribution of passing rules. The percentage of passing rules is around 2–5%, so neither common nor particularly rare. There are large portions of the rule space that do not produce any events, and these regions are correlated but not identical between microstructures. To see how much commonality there is, we can plot the passing rules colored by how many graphs passed:

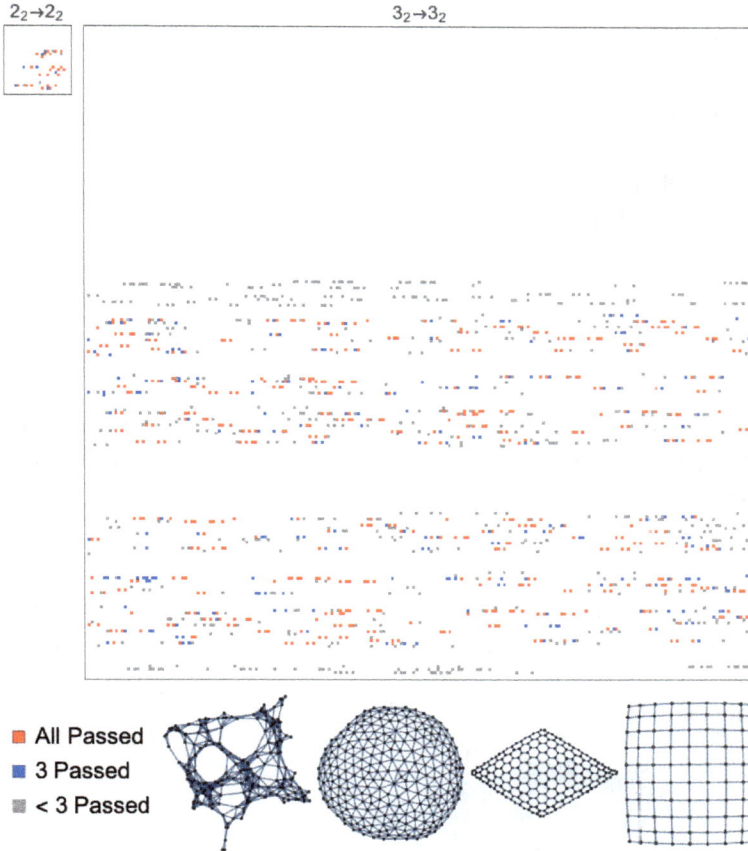

$2_2 \rightarrow 2_2$ $3_2 \rightarrow 3_2$

■ All Passed
■ 3 Passed
▨ < 3 Passed

Compute the percentage of rules that pass with one, two or all three of the input graphs:

In[·]:= **ConsistencyStats[ds][All, All, PercentForm]**

Out[·]=

$\{\{2, 2\}\} \rightarrow \{\{2, 2\}\}$	1		8.824%
	2		2.941%
	3		20.59%
	4		67.65%
$\{\{3, 2\}\} \rightarrow \{\{3, 2\}\}$	1		20.65%
	2		25.85%
	3		16.07%
	4		37.42%

We can see that 68% of rules with the signature $2_2 \rightarrow 2_2$ that pass for one of the inputs passes for all four inputs, and 37% for rules with signature $3_2 \rightarrow 3_2$. This shows us that although the rules are sensitive to microstructure, there is still a large percentage of rules that pass for all the microstructures tested.

Schwarzschild Graph

We would like to test the behavior of the rules with graphs corresponding to the Schwarzschild solution. To properly capture the geometric structure of the Schwarzschild solution, it is necessary to use a graph with more vertices than used previously (the flat graphs had around two hundred vertices). This proves to be computationally challenging, so to constrain the search space, we will only use the rules that passed for at least three out of four input graphs in the previous section. This leaves us with 719 rules to search over.

First, we filter the rules by those that passed four generations without failing any criteria:

```
In[ ]:=  winners =
            ds[Select[#["Termination Criteria"] == {} && #Generation ≥ 4 &]][GroupBy["Rule"],
                CountDistinct, "Input Graph Hash"][Select[# ≥ 3 &]] // Normal // Keys;
         Length[winners]
```

```
Out[ ]=  719
```

We repeat the search over just the passing rules from before, using a Schwarzschild graph containing five hundred vertices. Note that due to the limitations of the graph generation functions, this graph represents a two-dimensional spatial slice of the Schwarzschild solution, not a full three-dimensional solution:

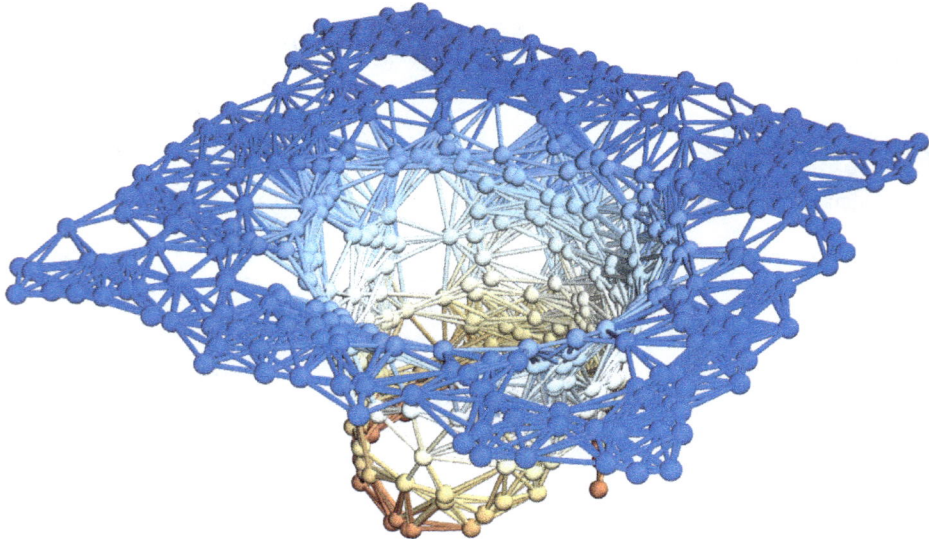

Each of the passing rules is then applied to the Schwarzschild graph for four generations and the heuristic criteria are evaluated.

This is a sample of the various behaviors of the Schwarzschild graph under the rules:

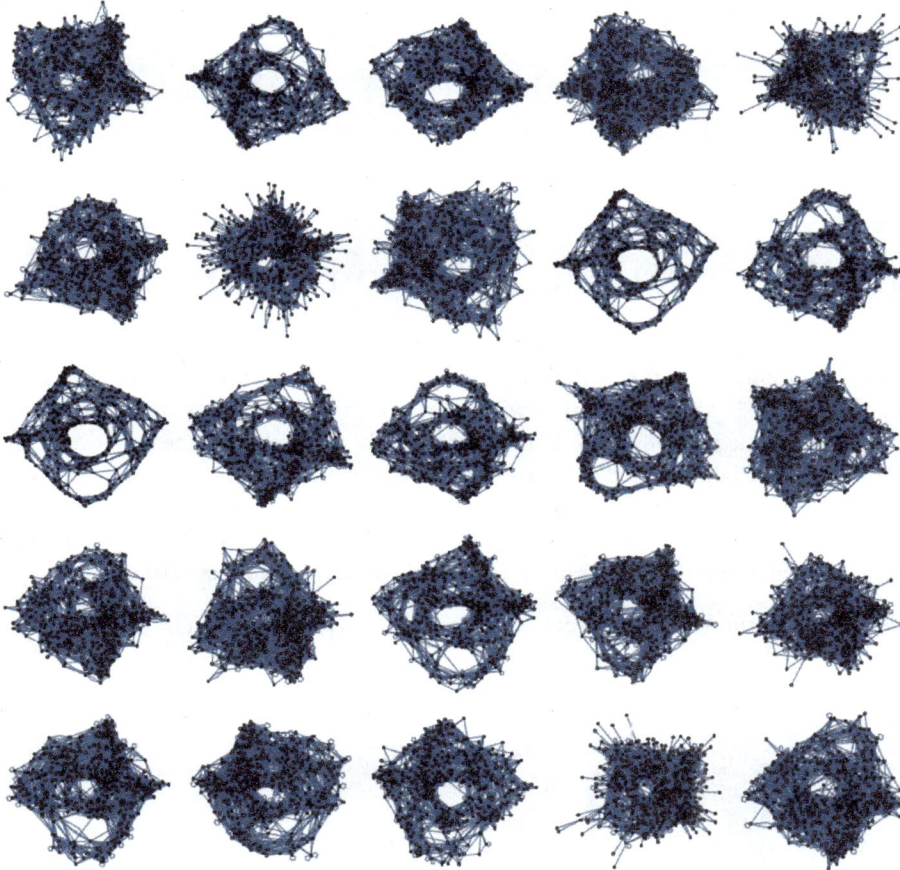

Now we can again analyze the termination criteria:

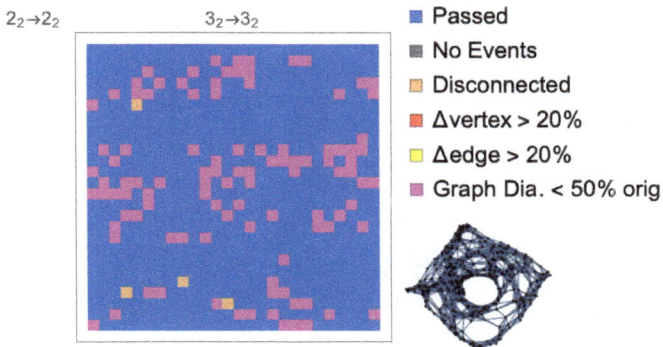

$2_2 \rightarrow 2_2$ $3_2 \rightarrow 3_2$

- ■ Passed
- ■ No Events
- ■ Disconnected
- ■ Δvertex > 20%
- ■ Δedge > 20%
- ■ Graph Dia. < 50% orig

In[ ]:= **TermStatsTable[dsDownselect]**

Out[ ]=

{{3, 2}} → {{3, 2}}	Disconnected	0.5806%
	Graph Dia. < 50% orig	15.67%
	None	83.74%
{{2, 2}} → {{2, 2}}	None	100%

We can see that the majority of rules continue to pass with the Schwarzschild graph; however, there are about 15% that fail.

Excluding Trivial Rules via Isomorphism

By visual inspection of the previous sample, we can see some resulting graphs that appear unchanged by application of the rule. We know from the number of events that the rules do, in fact, apply, but it is possible that they will result in a "trivial" transformation, for example swapping the directions of all edges. We can test for this by checking if the resulting graph is isomorphic to the original input. These rules may display physical behavior in more complex systems (e.g. in the presence of dynamics or "particles"); however, for the purposes of this study, we cannot evaluate them further. By excluding these rules, we can determine if there are nontrivial rules that also satisfy our criteria of geometry preservation.

The following plot shows the rules that result in an isomorphic graph highlighted in cyan:

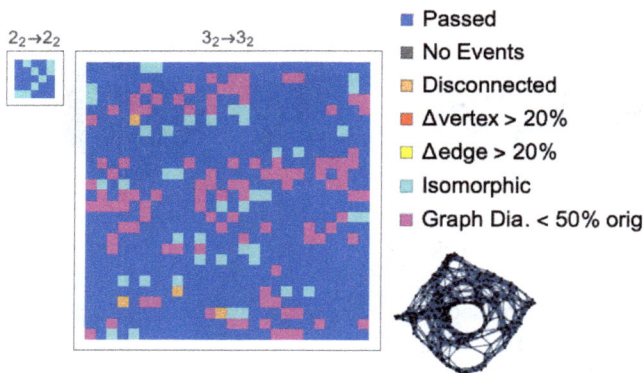

- Passed
- No Events
- Disconnected
- Δvertex > 20%
- Δedge > 20%
- Isomorphic
- Graph Dia. < 50% orig

In[ ]:= **TermStatsTable[dsDownselectIso]**

Out[ ]=

{{3, 2}} → {{3, 2}}	Disconnected	0.5806%
	Graph Dia. < 50% orig	15.67%
	Isomorphic	6.386%
	None	77.36%
{{2, 2}} → {{2, 2}}	Isomorphic	33.33%
	None	66.67%

We can see that about 6% of rules with three edges and 33% of rules with two edges result in isomorphic graphs.

Dimension Change under Rules

Excluding rules that failed on the Schwarzschild graph as well as the trivial rules, we are left with 551 rules remaining. One property that our rules should have is to preserve the approximate curvature and dimension of the manifold. As our input graph is still only five hundred vertices, curvature estimates are "noisy"; however, we can look more carefully at the change in dimension.

Our original criteria included a crude threshold on graph diameter that together with the threshold on vertex count acts to exclude rules that cause large changes in dimension, but this is not a very sensitive test. With our smaller search space, we can instead plot the $\Delta(r)$ function [1] to get a finer-grained view into how the dimension of the graph is affected by the rules.

In this plot, each blue line corresponds to the resulting graph from a different rule. Each line shows $\Delta(r)$ as a function of r, averaged over five starting vertices around the graph center. The red line shows $\Delta(r)$ for the initial input graph. Note that it has a plateau around two as expected, as the Schwarzschild solution was prepared as a 2D slice due to the limitations of the graph generation function (generalizing to 3D is a topic for future work):

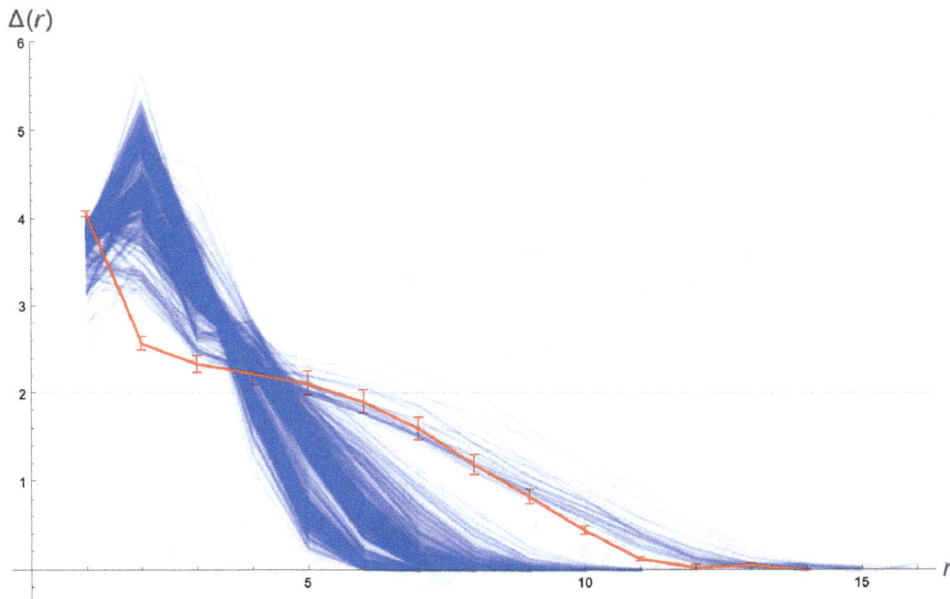

The plot shows that the rules can be grouped into two different behaviors. One group has $\Delta(r)$ spiking up at small values of r before quickly trending to zero. This corresponds to graphs that have become densely connected corresponding to higher dimension. There is a second group of rules that follows more closely the original input graph behavior. To discriminate between these two groups, it is sufficient to threshold on graph diameter >13. These rules are now highlighted in orange:

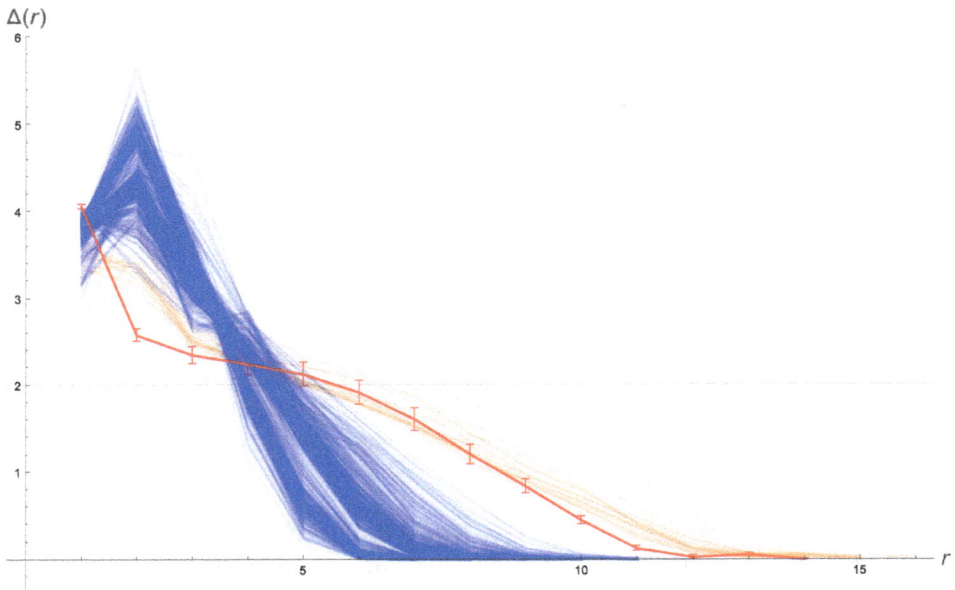

Applying this threshold, we are left with just 32 rules.

Results

To recap, we first did an exhaustive search over rules involving two and three edges using four small input graphs corresponding to flat two-dimensional space. The graphs were evolved according to each of the heuristic criteria for four generations, and those rules that passed all criteria for at least three out of four graphs were selected. These rules were then tested again on a larger Schwarzschild graph and filtered according to the criteria. Additional criteria for isometry and dimension preservation were applied, resulting in 32 remaining rules. The 32 remaining rules are shown here together with the corresponding final state of the Schwarzschild graph after four generations:

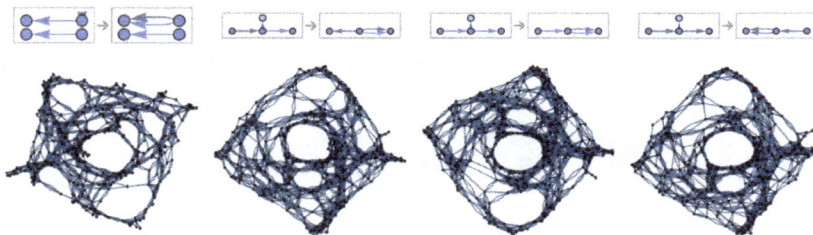

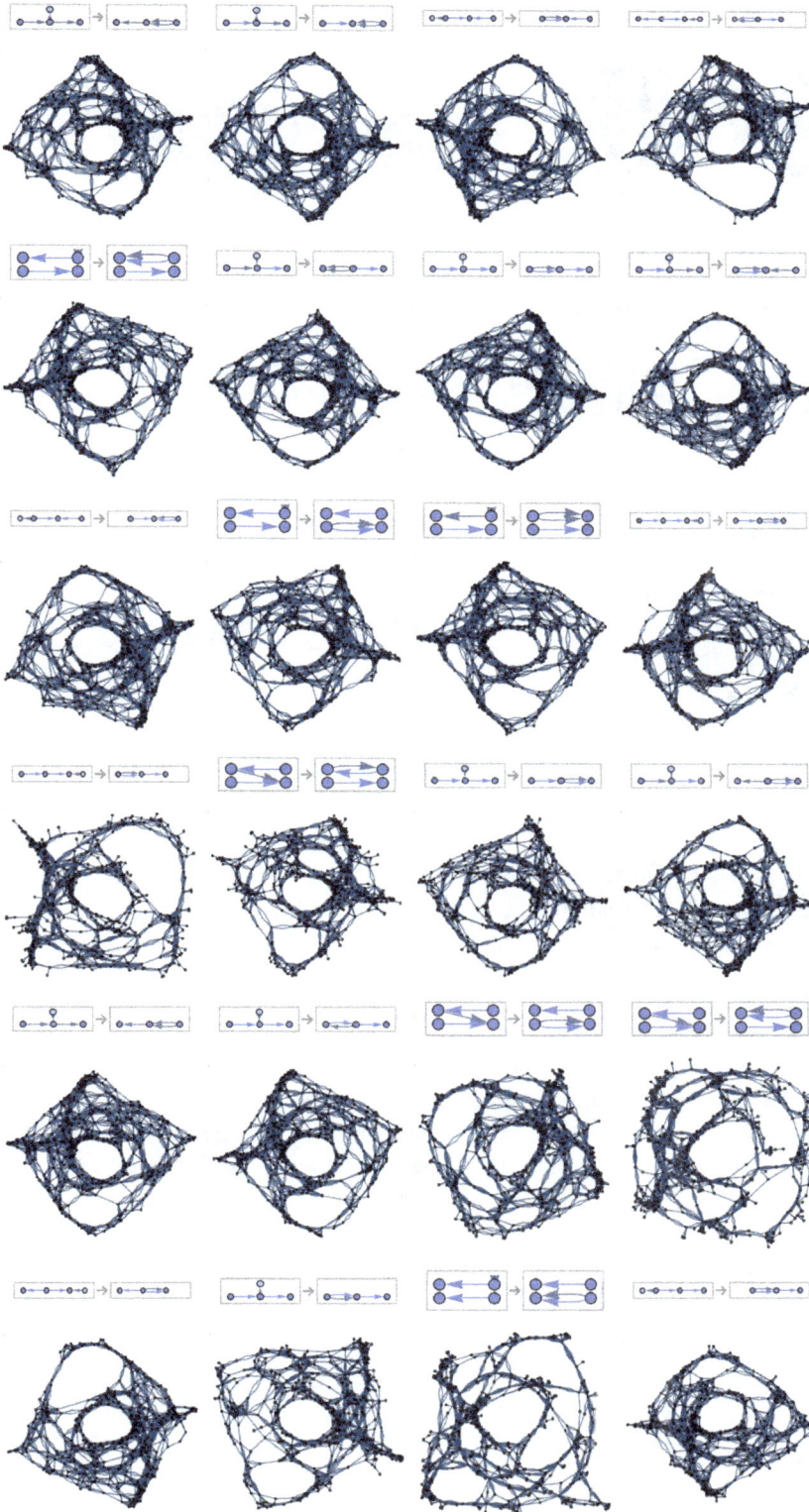

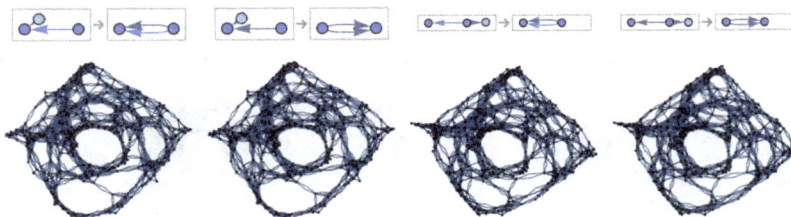

The numbers of remaining rules per signature were:

In[ ]:= **dsDownselectDimThres[GroupBy["Rule Signature"], CountDistinct, "Rule"]**

Out[ ]=

{{3, 2}} → {{3, 2}}	28
{{2, 2}} → {{2, 2}}	4

Full Set of Rules

Visit wolfr.am/WSS2024-Noble.

Concluding Remarks

In this project, I presented a method to perform a systematic search for Wolfram model rewriting rules that preserve some notions of the geometric structure of the input graph. I also introduced a set of heuristic rules to evaluate the rules for physicality and geometry preservation. By searching for rules with signatures $2_2 \to 2_2$ and $3_2 \to 3_2$ on simple directed graph inputs corresponding to flat Minkowski and Schwarzschild solutions of the Einstein equations, I identified 32 rules out of 49,116 that preserve the geometry of space according to the heuristic criteria.

To enable this systematic search, several simplifying assumptions were made. Small graph "patches" were considered, assuming boundary effects were not significant. We also assumed the rules would include the same number of input and output edges, potentially overlooking suitable rules that do not meet this condition. To make the search tractable, we limited the number of edges in the rule to three. Although there is reason to believe that rules involving four edges could exhibit more complex behavior [1], computational constraints prevented a systematic search over these rules. The heuristics selected provided an initial test of geometry preservation, but did not fully evaluate the geometry, such as full curvature. Additionally, due to the limitations of the DiscreteHypersurfaceDecomposition function, we were only able to test a two-dimensional slice of the Schwarzschild solution.

In summary, we can draw the following conclusions:

- Systematic search using heuristic criteria can be a powerful tool for exploring the space of rules in the Wolfram model.

- Suitable rules that preserve geometry are relatively rare for the cases we considered, comprising just 0.07% of rules evaluated, but several candidate rules were found.

- Sensitivity to microstructure is a significant challenge for systematic rule searches, as we do not know what the correct vacuum microstructure(s) will be in Wolfram models that correspond to physical behavior.

Future Work

This project attempted to provide the simplest possible version of a systematic rule search, and there are many possible avenues for future work:

- Extensions to larger graphs and longer evolutions

- Testing the full three-dimensional Schwarzschild solution

- Trying more complex vacuum solutions, such as the Kerr metric, that include shear

- Using more sophisticated heuristics such as curvature measures (e.g. Ollivier–Ricci curvature)

- Testing if these rules produce the expected dynamics in nonstatic or nonvacuum scenarios

- It may be possible to reason about the sensitivity to microstructure by finding equivalencies between classes of rules and microstructures

Use the Rules in Your Project

If you'd like to experiment with the rules from this project, they are available from the Wolfram Data Repository:

```
In[ ]:=  rules = ResourceData[ResourceObject[CloudObject[
             "https://www.wolframcloud.com/obj/fergusnoble/DeployedResources/Data/
             Geometry–Preserving–Wolfram–Model–Rules"]]]
```

Acknowledgments

I would like to thank my mentors, Nik Murzin and Xerxes D. Arsiwalla, for their guidance and support. I also thank Gianmarco Morbelli and Jon Lederman for the valuable discussions that contributed to this project. I would also like to extend my special thanks to Stephen Wolfram for suggesting this project, and to all of the Wolfram Summer School team for putting together this fantastic program.

References

1. S. Wolfram (2020), "A Class of Models with the Potential to Represent Fundamental Physics," *The Wolfram Physics Project*. www.wolframphysics.org/technical-introduction.

2. J. Gorard (2021), "Some Relativistic and Gravitational Properties of the Wolfram Model," arXiv:2004.14810v2.

Access the Full Code

Scan or visit wolfr.am/WSS2024-Noble.

Cite This Notebook

"Search for Rules that Preserve Geometric Structure in Wolfram Models"
by Fergus Noble
Wolfram Community, STAFF PICKS, July 9, 2024
community.wolfram.com/groups/-/m/t/3210162

Robotic Arm Control Using Large Language Models

CURRAN FLANDERS

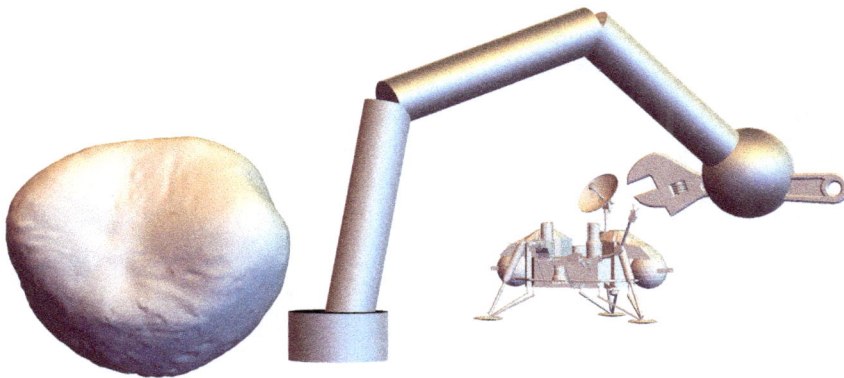

The use of large language models (LLMs) can add natural language comprehension and reasoning abilities to a robot control system. This project uses the capabilities of LLMs to enable a robotic arm to follow natural language instructions. The arm is simulated in Mathematica in an environment containing other objects. First, a 2D model of the arm is constructed. Then the arm is extended into 3D. Functionality is added to allow the arm to be used in different environments. Each environment contains a collection of 3D objects. Finally, functionality is added to allow the LLM to guide the movement of the arm by specifying a list of sequential target positions.

Implementing an LLM-Controlled Robot Arm in Mathematica

The purpose of this project is to create a workflow and code that allow a user to control a simulated robot arm with natural language instructions. The natural language instructions will be passed to a large language model (LLM), which will interpret the instructions and determine the positions to which the arm will be moved. The LLM specifies points in space for the arm to move to, but the inverse kinematics must be handled in a separate process.

The inputs to the LLM are a prompt from the user, a string description of the arm's environment and a template string to enclose this data. The LLM returns a path of points whose trajectory the arm should follow. Inverse kinematics is used to determine the joint angles the robot arm must have to reach these positions. An interpolation function is created between the sets of joint angles corresponding to the points along the path. This leads to a continuous path between the sets of points. A graphic showing the arm traveling along its path is displayed.

This process should work in a similar manner for any LLM whose API is directly accessible by Wolfram, but testing has been performed exclusively with GPT-4 by OpenAI.

2D Kinematics

The 2D model of the robot arm, which we will extend to the 3D case, has two degrees of freedom. The angle between the base and the arm's first segment will be referred to as $\theta 1$. The angle between the first and second segment and the angle between the second and third segment are identical and referred to as $\theta 2$. If the length of each segment is 1, the position of the hand in terms of the joint angles is as follows:

$x = (1 + 2 \cos(\theta 2)) \cos(\theta 1 + \theta 2)$
$y = (1 + 2 \cos(\theta 2)) \sin(\theta 1 + \theta 2)$

The analysis of robotic arms involves both forward kinematics and inverse kinematics. A robot arm has a number of degrees of freedom, each of which is a variable that typically specifies an arm segment's angular position around an axis or the angle that it forms with another segment. These degrees of freedom can be specified as a set of angles that together completely describe the arm's position at a given time. Forward kinematics is the process of using these joint angles to find the position of the robot's hand. Inverse kinematics is the opposite procedure, finding the joint angles from the position of the hand. Although forward kinematics is a function such that every input is mapped to a unique output, inverse kinematics is not a function, since there may be multiple possible sets of angles that map to a given position.

2D Forward Kinematics

Determine the point in 2D that the robot's hand reaches when its arm joints are bent at specified angles:

```
In[•]:=  ClearAll[robotArmForwardKinematics2D]
         robotArmForwardKinematics2D[θ1_?NumericQ, θ2_?NumericQ] :=
           With[{base = {0, 0}},

             RotationTransform[θ1, base][{1, 0}] + RotationTransform[θ1 + θ2, base][{1, 0}] +
               RotationTransform[θ1 + 2 θ2, base][{1, 0}]]
```

2D Inverse Kinematics

Determine the angles at which the arm's joints must bend in order for its hand to reach a specific point specified with 2D Cartesian coordinates:

In[ ]:= **ClearAll[robotArmInverseKinematics2D]**

robotArmInverseKinematics2D[{x_ ? NumericQ, y_ ? NumericQ}] := (NSolve[{
    RotationTransform[θ1, {0, 0}][{1, 0}] + RotationTransform[θ1 + θ2, {0, 0}][{1, 0}] +
        RotationTransform[θ1 + 2 θ2, {0, 0}][{1, 0}] == {x, y},
    0 ≤ θ1 ≤ π, −π/2 ≤ θ2 ≤ 0}, {θ1, θ2}] //
    If[Length[♯] > 0 && y ≥ 0, Values[First[♯]], Failure["PositionUnreachable",
        "MessageTemplate" → "This point is unreachable."]] &)

2D Graphics

Make a visual representation of the 2D cross-section of the arm at specified joint angles:

In[ ]:= **ClearAll[robotArmAnglePlot2D]**

robotArmAnglePlot2D[{θ1_, θ2_}] := Module[{base = {0, 0}, elbow, elbow2, hand},
    elbow = RotationTransform[θ1, base][{1, 0}];
        elbow2 = elbow + RotationTransform[θ1 + θ2, base][{1, 0}];
        hand = elbow2 + RotationTransform[θ1 + 2 θ2, base][{1, 0}];
        Graphics[{Circle[], Line[{base − {2, 0}, base + {2, 0}}],
            Arrow[{base, elbow}], Arrow[{elbow, elbow2}], Arrow[{elbow2, hand}]}]];

In[ ]:= **Row[{robotArmAnglePlot2D[{0, 0}], robotArmAnglePlot2D[{π/3, −π/4}]}]**

Out[ ]=

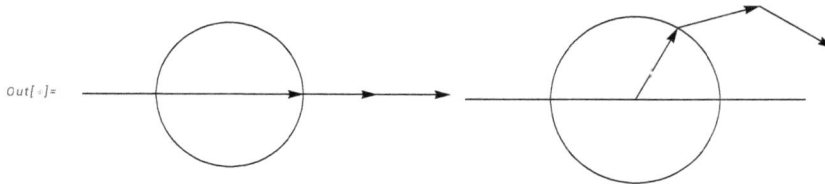

Make a visual representation of the 2D cross-section of the arm with joint angles automatically calculated such that the hand reaches a specified set of (*x*, *y*) coordinates:

In[ ]:= **ClearAll[robotArmPositionPlot2D]**

robotArmPositionPlot2D[handCoords_List] :=
    Module[{θ1, θ2, base = {0, 0}, elbow, elbow2, hand}, If[FailureQ[♯], ♯,
        {θ1, θ2} = ♯;
    elbow = RotationTransform[θ1, base][{1, 0}];
            elbow2 = elbow + RotationTransform[θ1 + θ2, base][{1, 0}];
            hand = elbow2 + RotationTransform[θ1 + 2 θ2, base][{1, 0}];
            Graphics[{Circle[], Line[{base − {2, 0}, base + {2, 0}}], Arrow[
                {base, elbow}], Arrow[{elbow, elbow2}], Arrow[{elbow2, hand}]}]] &@
        robotArmInverseKinematics2D[handCoords]];

3D Kinematics and Graphics

To extend this model into the third dimension, we define an additional angle, swivelθ. This variable represents the rotation of the arm around the axis that is perpendicular to the arm's base and passes through the center of the base. The 3D model can be visualized as follows:

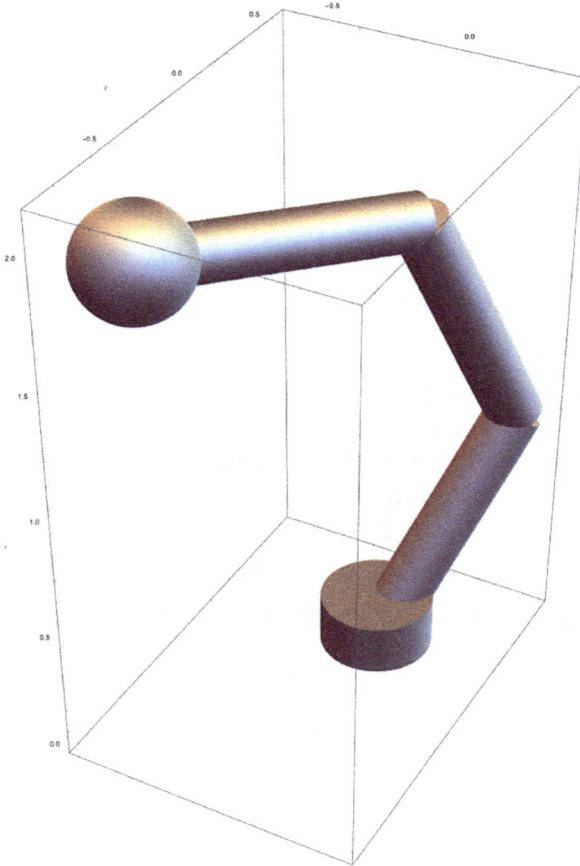

This leads us to the 3D extension of the equations described in the section on 2D kinematics:

$x = (1 + 2 \cos(\theta 2)) (\cos(\theta 1 + \theta 2)) \cos(\text{swivel}\theta)$
$y = (1 + 2 \cos(\theta 2)) (\cos(\theta 1 + \theta 2)) \sin(\text{swivel}\theta)$
$z = (1 + 2 \cos(\theta 2)) \sin(\theta 1 + \theta 2)$

3D Forward Kinematics

Determine the location of the robot's hand given its joint angles:

```
In[ ]:=  ClearAll[robotArmForwardKinematics3D]
         robotArmForwardKinematics3D[swivelθ_ ? NumericQ,
           θ1_ ? NumericQ, θ2_ ? NumericQ] := RotationTransform[swivelθ, {0, 0, 1}][
           Insert[robotArmForwardKinematics2D[θ1, θ2], 0, 2]]
```

3D Inverse Kinematics

Determine a set of joint angles that will cause the robot's hand to reach the specified position:

```
In[ ]:= ClearAll[robotArmInverseKinematics3D]
        robotArmInverseKinematics3D[{x_ ? NumericQ, y_ ? NumericQ, z_ ? NumericQ}] :=
          Module[{swivelθ},
              swivelθ = If[x > 0, ArcTan[y / x], ArcTan[y / x] + π];
                swivelθ = Mod[swivelθ, 2 π, -π];

              If[FailureQ[#], #, Prepend[#, swivelθ]] & @ robotArmInverseKinematics2D @
                Delete[RotationTransform[-swivelθ, {0, 0, 1}][{x, y, z}], 2]
                ]
```

3D Graphics

Plot the robot position in 3D given angle specifications:

```
In[ ]:= ClearAll[robotArmAnglePlot3D]
        Options[robotArmAnglePlot3D] = {Axes → True, AxesLabel → {"x", "y", "z"},
            PlotRange → 3.2 {{-1, 1}, {-1, 1}, {-.01, 1}}, ImageSize → Medium};
        robotArmAnglePlot3D[swivelθ_ : 0, θ1_ : 0, θ2_ : 0, OptionsPattern[]] :=
          Module[{base = {0, 0, 0}, swivelVector, swivelVectorNormal, elbow, elbow2, hand},
                swivelVector = Append[AngleVector[swivelθ], 0];
                                                          π
                swivelVectorNormal = Append[AngleVector[swivelθ - ─], 0];
                                                          2
                elbow = RotationTransform[θ1, swivelVectorNormal][swivelVector];

            elbow2 = elbow + RotationTransform[θ1 + θ2, swivelVectorNormal][swivelVector];

            hand = elbow2 + RotationTransform[θ1 + 2 θ2, swivelVectorNormal][swivelVector];
                Graphics3D[{
                            (*Robot base: (lazy zusan)*)
                            FaceForm[Opacity[.1]], Cylinder[{base, base + {0, 0, .2}}, 1],
                    Dashed, Arrow[{base, swivelVector}], Dashing[None],
                            (*Robot upper arm:*)
                            Blue, Cylinder[{base, elbow}, .1],
                            (*Robot middle arm:*)
                            Cylinder[{elbow, elbow2}, .1],
                            (*Robot lower arm:*)
                            Cylinder[{elbow2, hand}, .1],
                            (*Robot hand:*)
                            Ball[hand, .2]},
                      Axes → OptionValue[Axes], AxesLabel → OptionValue[AxesLabel],
                      PlotRange → OptionValue[PlotRange],
                ImageSize → OptionValue[ImageSize]]]
```

In[ ]:= **robotArmAnglePlot3D @@**

    robotArmInverseKinematics3D @ First[randomRobotGoalPositions[1]]

Out[ ]=

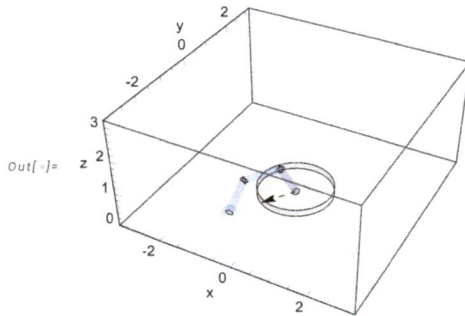

Create a simple 3D graphic of the robot arm with the hand at specified Cartesian coordinates:

In[ ]:= **ClearAll[robotArmPositionPlot3D]**

    Options[robotArmPositionPlot3D] = {Axes → True, AxesLabel → {"x", "y", "z"},

        PlotRange → 3.2 {{-1, 1}, {-1, 1}, {-.01, 1}}, ImageSize → Medium};

    robotArmPositionPlot3D[handCoords_List, OptionsPattern[]] :=

    Module[{swivelθ, θ1, θ2, base = {0, 0, 0}, swivelVector,

        swivelVectorNormal, elbow, elbow2, hand}, If[FailureQ[#], #,

      {swivelθ, θ1, θ2} = #;

      swivelVector = Append[AngleVector[swivelθ], 0];

          swivelVectorNormal = Append[AngleVector[swivelθ − $\frac{\pi}{2}$], 0];

        elbow = RotationTransform[θ1, swivelVectorNormal][swivelVector];

        elbow2 =

    elbow + RotationTransform[θ1 + θ2, swivelVectorNormal][swivelVector];

        hand =

    elbow2 + RotationTransform[θ1 + 2 θ2, swivelVectorNormal][swivelVector];

        Graphics3D[{

      (∗Robot base: (lazy zusan)∗)

              FaceForm[Opacity[.1]], Cylinder[{base, base + {0, 0, .2}},

      1], Dashed, Arrow[{base, swivelVector}], Dashing[None],

          (∗Robot upper arm:∗)

          Blue, Cylinder[{base, elbow}, .1],

          (∗Robot middle arm:∗)

          Cylinder[{elbow, elbow2}, .1],

          (∗Robot lower arm:∗)

          Cylinder[{elbow2, hand}, .1],

          (∗Robot hand:∗)

          Ball[hand, .2]},

        Axes → OptionValue[Axes], AxesLabel → OptionValue[AxesLabel],

          PlotRange → OptionValue[PlotRange],

         ImageSize → OptionValue[ImageSize]]]] & @

      robotArmInverseKinematics3D[handCoords]

This shows the robot arm moving to a valid position:

```
In[•]:= robotArmPositionPlot3D[{2.0391310416650805`, 0.`, 0.41335232586276466`}]
```

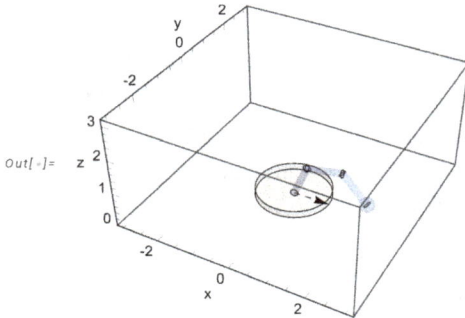

```
Out[•]=
```

This demonstrates how errors are caught if an invalid position is entered:

```
In[•]:= (*The angle in the argument is not reachable*)
       robotArmPositionPlot3D[{2.0391310416650805`, 0.`, -1}, PlotRange → All]
```

```
Out[•]= Failure[    ⚠   Message:   This point is unreachable.
                       Tag:       PositionUnreachable
                                                              ]
```

Plotting a Path for the Arm in 3D

Convert a discrete list of target positions, specified by the arm's joint angles, to a continuous function that outputs joint angles for any point along the arm's path:

```
In[•]:= ClearAll[getRobotArmMovementFunction]
       (*The function takes a list of lists of {{swivelθ,θ1,θ2},{...},...}*)
       getRobotArmMovementFunction[angleSpecs_List] :=
           Interpolation[#, InterpolationOrder → 1] &@
               MapIndexed[{First[#2], #1} &, angleSpecs]
```

Generate a list of random 3D points on the ground that are reachable by the arm:

```
In[•]:= ClearAll[randomRobotGoalPositions]
       randomRobotGoalPositions[n_Integer] :=
               RotationTransform[RandomReal[{-π, π}], {0, 0, 1}] /@
           (Append[#, 0] &/@ RandomPoint[Annulus[{0, 0}, {1, 3}], n])
```

Defining the Environment

We will define each environment for the arm as a dataset that includes three columns: position, descriptions and geometry. The LLM will receive the position and description of each object from the prompt. The geometry will be used to visualize the object in 3D.

Creating Environment Datasets

Generate roughly equally spaced positions for objects in the environment:

```
In[ ]:=  ClearAll[precomputedPositions]
         precomputedPositions[n_Integer] := With[{swivelθ = RandomReal[{−π, π}]},
             RotationTransform[swivelθ, {0, 0, 1}] /@
               (Append[#, 0] & /@ (RandomPoint[Annulus[{0, 0}, {1, 3}], n]))] /; n ≤ 3
         precomputedPositions[n_Integer] := With[{swivelθ = RandomReal[{−π, π}]},
             RotationTransform[swivelθ, {0, 0, 1}] /@ (Append[#, 0] & /@ ( {…} ⟦n−3⟧))] /;
           3 < n ≤ 26
```

```
In[ ]:=  Graphics3D[Point[precomputedPositions[26]], ImageSize → Tiny]
```

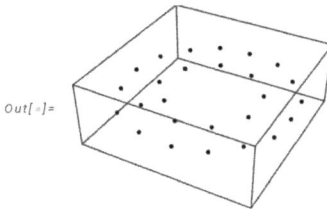

Out[ ]=

Helper function: translate 3D objects:

```
In[ ]:=  ClearAll[translate3DObject]
         translate3DObject[object_Graphics3D, vect_List:{0, 0, 0}] := First[MapAt[
               TranslationTransform[vect], object,
               Map[Append[#, 1] &, Position[object, _GraphicsComplex]]]]

         translate3DObject[object_List, vect_List:{0, 0, 0}] :=
           translate3DObject[Graphics3D[object], vect]
```

Helper function: scale 3D objects:

```
In[ ]:=  ClearAll[scale3DObject]
         scale3DObject[object_Graphics3D, s_:1, p_List:{0, 0, 0}] := First[MapAt[
               ScalingTransform[{s, s, s}, p], object,
               Map[Join[#, {1}] &, Position[object, _GraphicsComplex]]]]

         scale3DObject[object_List, s_:1, p_List:{0, 0, 0}] :=
           scale3DObject[Graphics3D[object], s, p]
```

Get a 3D object's dimensions:

```
In[ ]:=  ClearAll[get3DObjectDimensions]
         get3DObjectDimensions[object_] :=
           Abs[Subtract @@@ Transpose[CoordinateBoundingBox[
               Flatten[Cases[object, g_GraphicsComplex :→ First[g], ∞], 1]]]]
```

Get the centroid of the base of a 3D object:

```
In[ ]:=  ClearAll[get3DObjectBaseCentroid]
         get3DObjectBaseCentroid[object_] := With[{bbox = CoordinateBoundingBox[
               Flatten[Cases[object, g_GraphicsComplex :> First[g], ∞], 1]]},
               Append[Mean[bbox[[All, {1, 2}]]], bbox[[1, 3]]]]
```

Center the base of a 3D object at the origin and scale it so its largest dimension is equal to length:

```
In[ ]:=  ClearAll[standardize3DObject]
         standardize3DObject[object_, length_ : 1] :=
           (translate3DObject[#, –get3DObjectBaseCentroid[#]] & @
               scale3DObject[object, length / Max[get3DObjectDimensions[object]]])
```

An example environment is generated here. It uses objects from the built-in "Geometry3D" data.

Define some example objects to include in the robot's environment:

```
In[ ]:=  ClearAll[standardizedExampleObjects]
         standardizedExampleObjects = Graphics3D[#, Axes → True] & /@
               standardize3DObject /@ ExampleData /@ ExampleData["Geometry3D"];
```

Associate each object with a description:

```
In[ ]:=  objectsWithDescriptions = Thread[{
               {"Bass Guitar", "Beethoven", "Castle Wall",
                "Cone", "Cow", "Deimos", "Galleon", "Hammerhead Shark",
                   "Horse", "Klein Bottle", "Moebius Strip",
                "Phobos", "Potted Plant", "Seashell", "Sedan Car",
                   "Space Shuttle", "Stanford Bunny",
                "Torus", "Tree", "Triceratops", "Tugboat", "Utah Teapot",
                   "Utah VW Bug", "Vase", "Viking Lander", "Wrench", "Zeppelin"},
                   standardizedExampleObjects}];
```

Generate an environment with *n* objects for the LLM + arm system:

```
In[ ]:=  ClearAll[generateArmEnvironment]
         generateArmEnvironment[n_] := Module[{pos = precomputedPositions[n],
               obj = RandomSample[objectsWithDescriptions, n], env},
               env = Transpose @ Dataset[AssociationThread[
                   {"Position", "Description", "Geometry"},
                   Transpose[Join @@@ Thread[{List /@ pos, obj}]]]];
               env[All, <|#,
                   "Geometry" → Graphics3D @ translate3DObject[#Geometry, #Position]|> &]]
```

In[ ]:= **exampleEnv = generateArmEnvironment[4]**

Out[ ]=

Position	Description	Geometry
{1.80532, 0.728394, 0.}	Tugboat	...
{−1.80639, −0.729872, 0.}	Potted Plant	...
{0.726987, −1.81361, 0.}	Cow	...
{−0.734835, 1.80921, 0.}	Horse	...

Accessing Data from an Environment Dataset

Retrieve geometry from the environment for visualizations:

In[ ]:= **ClearAll[environmentGeometry]**
environmentGeometry[env_Dataset] := First/@Normal[env[All, "Geometry"]]

Retrieve information from the environment for the LLM:

In[ ]:= **ClearAll[describeEnvironment]**
describeEnvironment[env_Dataset] := Normal[env[All, {"Position", "Description"}]]

Graphing the Robot Arm in 3D with an Environment

Plot the robot arm in its environment:

In[ ]:= **ClearAll[robotArmPlot3D]**
Options[robotArmPlot3D]={PlotRange→3.2{{−1,1},{−1,1},{−.01,1}}};
robotArmPlot3D[handAngles_List:{1,1,1},environment_Dataset:Dataset[{}]] :=
 Module[{swivelθ,θ1,θ2,base={0,0,0},
  swivelVector,swivelVectorNormal,elbow,elbow2,hand},
   {swivelθ,θ1,θ2}=handAngles;
   swivelVector=Append[AngleVector[swivelθ],0];

   $$\text{swivelVectorNormal=Append[AngleVector[swivel}\theta-\frac{\pi}{2}\text{],0];}$$

   elbow=RotationTransform[θ1,swivelVectorNormal][swivelVector];

   elbow2=elbow+RotationTransform[θ1+θ2,swivelVectorNormal][swivelVector];

   hand=elbow2+RotationTransform[θ1+2θ2,swivelVectorNormal][swivelVector];
    Graphics3D[{
     (*Robot base: (lazy zusan)*)
      FaceForm[Opacity[1]],
    {MaterialShading["Silver"],Cylinder[{base,base+{0,0,.2}},0.2],
       (*Robot upper arm:*)
       LightGray,Cylinder[{base,elbow},.1],
       (*Robot middle arm:*)

```
        Cylinder[{elbow,elbow2},.1],
        (*Robot lower arm:*)
        Cylinder[{elbow2,hand},.1],
        (*Robot hand:*)
        Ball[hand,.2]},
    environmentGeometry[environment]},
    Axes→True,
    AxesLabel→{"x","y","z"},PlotRange→All,ImageSize→500]]
```

In[ ]:= **robotArmPlot3D[{1, 1, 1}, exampleEnv]**

Out[ ]=

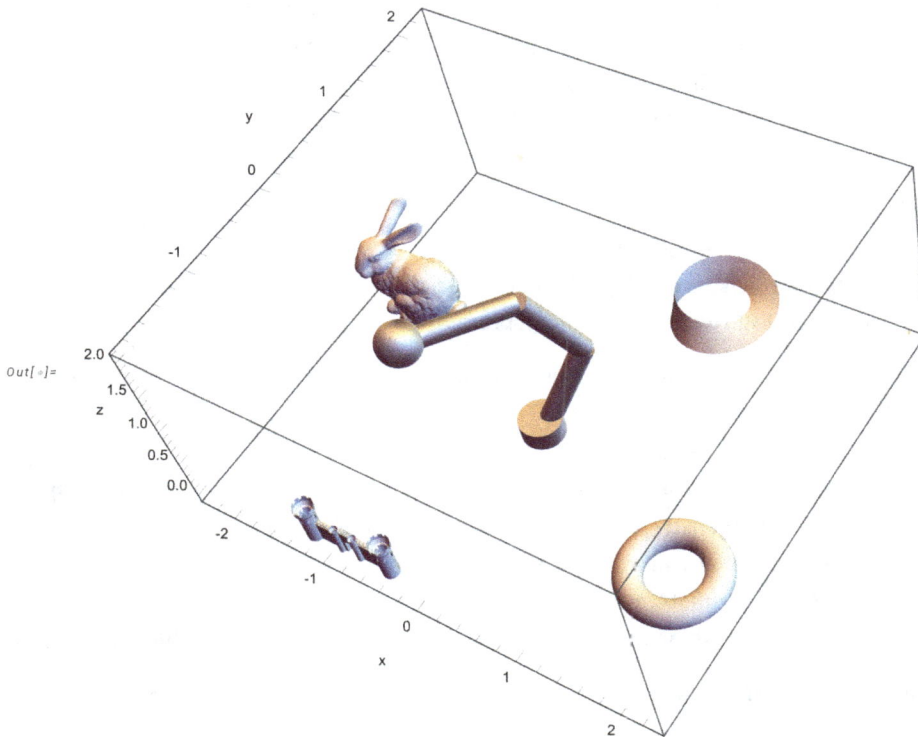

Controlling the Arm with an LLM

We desire for the LLM to provide coordinate points for the arm to travel to sequentially. The prompt uses an identical prompt template for each command. The template is modified by adding the environment of the arm and the user's instructions. However, the LLM cannot accept the environment objects directly as 3D graphics objects due to the maximum prompt size. For detailed objects, especially those imported from 3D object files, this might overwhelm the maximum number of tokens in the prompt and provide data that is not understandable to the LLM.

A template for a prompt that will instruct the LLM to return a list of points for the arm to move to in sequence:

```
In[ ]:=  templateLLMPrompt =
             LLMFunction["You are a control system tasked with instructing a robot
                 arm in a 3D environment. To instruct the arm, give it a list of
             successive {x,y,z} coordinates (where z is height) to perform the specified task which will
                 follow at the end of this instruction set. The list should be
             in the format {position1,position2,position3,...}. Example: {{1,2,3},{4,5,6}} . Do not wrap
                 the output in an additional list, there should only be a single
             list containing one sublist for each coordinate. Return only this list of coordinates and
                 nothing else.\n\nA description of your environment as well as your
             task instructions will follow:\n\nENVIRONMENT
                 DESCRIPTION:`1`\n\nTASK INSTRUCTIONS:`2` "];
```

Inform the LLM of the arm's environment and its natural language instructions, and return the raw LLM output:

```
In[ ]:=  ClearAll[getLLMControlList]
         getLLMControlList[prompt_, environment_] :=
            templateLLMPrompt[environment, prompt]
```

Create an animation of the arm's joint angles moving between a list of specified values, changing the position of the hand:

```
In[ ]:=  ClearAll[robotArmAnimate3D]
         robotArmAnimate3D[angles_List, environment_Dataset] := Module[{pathFunction},
                 If[Length[angles] == 1, robotArmPlot3D[angles〚1〛, environment],
                 pathFunction = getRobotArmMovementFunction[angles];
                 Animate[robotArmPlot3D[pathFunction[x], environment],
                 {x, 1, Length[angles]}, AnimationRepetitions → 1, AnimationRunning → False]]]
```

Simulate the robot arm following natural language instructions from the user in a specified environment:

```
In[ ]:=  ClearAll[simulateLLMCommands]
         simulateLLMCommands[prompt_String, environment_Dataset] :=
                 Module[{angles, pathFunction},
             angles = Riffle[#, #] &@@ (robotArmInverseKinematics3D /@ Prepend[ToExpression[
                         getLLMControlList[describeEnvironment[environment], prompt]],
             {1, 1, 1}])]
```

Demonstrations and Testing

GPT-4 was used for this project's tests.

The arm was successfully able to perform the most basic task, touching all objects in the environment:

```
In[ ]:=  robotArmAnimate3D[
            simulateLLMCommands["Touch all four objects", exampleEnv], exampleEnv]
```

However, when we requested that it select only a specific type of object, it initially failed to do so. An example of this follows:

```
In[ ]:=  robotArmAnimate3D[
            simulateLLMCommands["Touch the animals", exampleEnv], exampleEnv]
```

Making it clearer that objects not fitting the specification should be excluded from the analysis resolved this issue:

```
In[ ]:=  robotArmAnimate3D[
            simulateLLMCommands["Touch only the animals", exampleEnv], exampleEnv]
```

When prompted to select an object of a specific color, the arm was able to recognize the correct object despite color not being present in the graphics component of the simulation:

```
In[ ]:=  robotArmAnimate3D[
            simulateLLMCommands["Touch something green", exampleEnv], exampleEnv]
```

The LLM could generate its own prompts to simulate the user's commands. It favored generating prompts that commanded the arm to touch the objects in a specific sequence. The LLM's instructions to the arm were correct regardless of whether the environment was sparse (five objects) or relatively dense (15 objects):

```
In[ ]:=  {sparseEnv, mediumDensityEnv, denseEnv} =
            Table[generateArmEnvironment[i], {i, 5, 15, 5}];
         robotArmPlot3D[{1, 1, 1}, mediumDensityEnv
```

```
In[ ]:=  testPrompts =
            LLMFunction["Generate a set of instructions for a robot arm to perform based on
                    the following environment description. Do not tell
                    the robot arm to pick up or move any of the objects.
                    Only tell them to touch the objects according to
                    your choice of criteria. The environment description
                    follows. It contains object descriptions. Please
                    provide simple natural language instructions,
                    as you would to a human being. \n\n`1` ",
                 InsertionFunction → TextString][#] &/@
             (describeEnvironment[#]〚All, "Description"〛 &/@
                 {sparseEnv,
                    mediumDensityEnv,
                    denseEnv,
                    veryDenseEnv});
```

In[ ]:= **test1Instructions = testPrompts⟦1⟧**

Out[ ]= 1. Touch the Galleon.
2. Touch the Sedan Car.
3. Touch the Wrench.
4. Touch the Cone.
5. Touch the Bass Guitar.

In[ ]:= **test1Animation =**
robotArmAnimate3D[simulateLLMCommands[testPrompts⟦1⟧, sparseEnv], sparseEnv]

In[ ]:= **test2Instructions = testPrompts⟦2⟧**

Out[ ]= 1. Touch the Horse.
2. Touch the Seashell.
3. Touch the Cone.
4. Touch the Space Shuttle.
5. Touch the Beethoven statue.
6. Touch the Utah Teapot.
7. Touch the Vase.
8. Touch the Zeppelin.
9. Touch the Triceratops.
10. Touch the Castle Wall.

In[ ]:= **test2Animation = robotArmAnimate3D[**
simulateLLMCommands[testPrompts⟦2⟧, mediumDensityEnv], mediumDensityEnv]

In[ ]:= **test3Instructions = testPrompts⟦3⟧**

Out[ ]= 1. Touch the Beethoven statue.
2. Touch the Castle Wall.
3. Touch the Tugboat.
4. Touch the Zeppelin.
5. Touch Deimos.
6. Touch the Viking Lander.
7. Touch the Galleon.
8. Touch the Moebius Strip.
9. Touch the Tree.
10. Touch the Triceratops.
11. Touch the Stanford Bunny.
12. Touch the Wrench.
13. Touch the Bass Guitar.
14. Touch the Seashell.
15. Touch the Hammerhead Shark.

In[ ]:= **test3Animation =**
robotArmAnimate3D[simulateLLMCommands[testPrompts⟦3⟧, denseEnv], denseEnv]

Concluding Remarks

In this project, I have created a basic model of a robotic arm that can be controlled by an LLM. It can serve as a novel proof of concept of an LLM's ability to identify characteristics of objects and control a simulated physical system in response to natural language prompting.

The LLM was able to identify objects in the environment by their properties, such as color. However, it sometimes defaulted to touching all objects in the environment rather than the ones specified in the prompt unless a keyword such as "only" was used. It was able to perform instructions sequentially and functioned in both sparse environments where there were few objects in reach and dense environments where there were many. The testing process could be partially automated by allowing the LLM itself to invent prompts. However, its imagination in inventing prompts was somewhat limited in this test.

Future work for this project is likely to include a mechanism to allow the arm to move objects within its environment. More robust prompting of the LLM and handling of its output to allow the arm to perform actions other than moving to the locations of specified objects is another goal. Further experimentation on the LLM prompt templates could provide insight into which behaviors of the LLM are prompt specific and which are inherent to the model. Finally, implementing the LLM control pipeline in Wolfram System Modeler would allow for more realistic object dynamics.

Acknowledgments

I would like to thank the Wolfram Summer School staff and students who provided advice and assistance to me throughout the project:

- My mentors Phileas Dazeley-Gaist and Alejandra Ortiz for assistance throughout the project on planning and implementation.
- Glen Halley and Ankit Naik for offering assistance with System Modeler.
- Jofre Espigulé Pons for providing reference materials.
- Dr. Stephen Wolfram for high-level guidance on project direction.

References

1. Y. Yamada, et al. (2023). "Evaluating Spatial Understanding of Large Language Models." arXiv: 2310.14540.

2. F. Zeng, et al. (2023). "Large Language Models for Robotics: A Survey." arXiv: 2311.07226.

Access the Full Code

Scan or visit wolfr.am/WSS2024-Flanders.

Cite This Notebook

"Robotic Arm Control Using Large Language Models"
by Curran Flanders
Wolfram Community, STAFF PICKS, July 9, 2024
community.wolfram.com/groups/-/m/t/3211121

Exploring the Distributions of Shallow Multi-layer Perceptrons

TEO BENAROUS

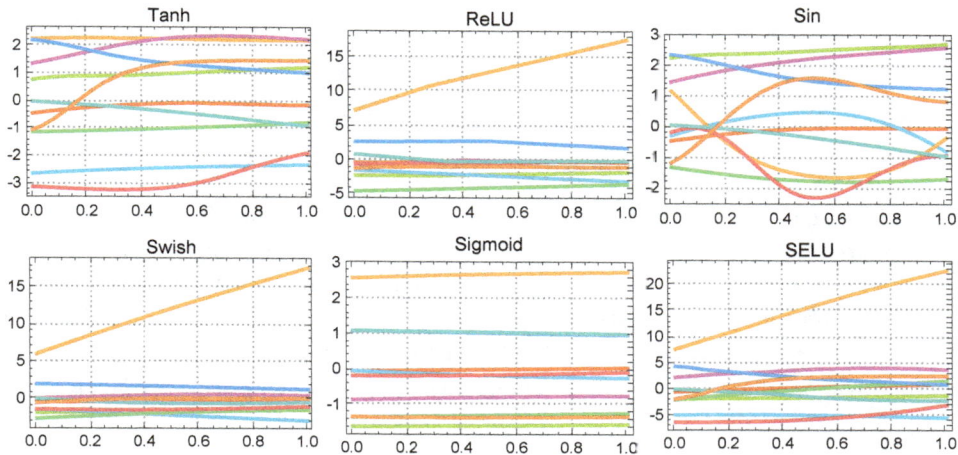

This work explores various statistical and mathematical techniques for analyzing and representing the distributions of shallow multi-layer perceptrons with two hidden layers and a width of five. The fixed architecture adheres to the universal approximation property as proven by [2]. To assess the potential of common activation functions, we have conducted a descriptive and comparative analysis of their functional spaces. We start by examining the basic statistics and additionally provide empirical estimates of the first four moments. A significant portion of the study focuses on the Karhunen–Loève expansion (KLE) and its application in deriving principal functional components from empirical covariance operator estimates. We analyze the approximation errors from truncated KLEs and investigate the learned distributions for smooth activations, highlighting typical functions and anomalies. The discrete wavelet transform (DWT) is employed to explore the spaces of networks induced by the non-smooth activation functions. We discuss coarse coefficients, the energy fraction of details and the main wavelet coefficients' projection to a lower-dimensional manifold. Additionally, we assess the learned distributions within this context.

Introduction

Project Description

We explore the distributions of random shallow multi-layer perceptrons (MLPs) with a fixed architecture. The collection of neural nets with random weights and biases sampled from independent standard normal distributions for a given activation function σ is a random variable on the space of continuous functions. More precisely, the fixed architecture has width 5 and depth 2, with real input and output. The purpose of this work is to get insights on the distribution for several activation functions and to compare them.

For a given activation function σ, the family of functions is of the form $x \mapsto w_3\, \sigma\left(W_2\, \sigma\left(w_1\, x + \beta_1\right) + \beta_2\right)$ where β_i and w_i are vectors of dimension 5 and W_2 is a 5×5 matrix. Hence, each realization inherits the continuity and the differentiability properties of σ and that collection is a linear subspace of continuous functions.

We study the following activation functions: Tanh, Swish, ReLU, Sigmoid, Sin and SELU.

Without loss of generality, we will restrict the study to functions on [0, 1]. The interval is discretized by steps of size 2^{-r} for a given refinement r (typically $r = 8$).

Theoretical Justification of the Project

A foundational result in neural network theory is the universal approximation theorem [1], which states that neural networks can approximate any continuous function on a compact domain. In particular, [2] algorithmically constructed a two-hidden-layer feedforward neural network with fixed weights and a smooth sigmoidal activation function that provides the universal approximation property. Their construction requires only $3d + 2$ hidden neurons to approximate any continuous function of d variables with arbitrary precision. In the univariate case, this reduces to a total of five hidden neurons.

- This result justifies the choice of our architecture that is capable of being dense in the space of continuous functions.

- To assess the potential of classic activation functions, we can conduct a descriptive and comparative analysis of their spaces of functions.

Agenda

We will follow these steps:

- Distributions of basic statistics
  - Mean, standard deviation, norm and range

- Empirical estimates of the first four moments functions
  - Mean, variance, skewness and kurtosis

- Karhunen–Loève expansion
  - Principal functional components from empirical covariance operator estimate
  - Similarity measures of generated networks
  - Analysis of approximation truncated expansion errors

- Learned distribution for smooth activations based on Karhunen–Loève expansion
  - Gaussian mixtures estimates
  - Typical functions and anomalies

- Discrete wavelet transform to explore the spaces of networks generated by ReLU and SELU
  - Coarse coefficients
  - Energy fraction of details
  - Feature space plot of the main wavelet coefficients
  - Gaussian mixture estimates

Setup

```
In[•]:=  width = 5;
         depth = 2;
         baseSeed = 123;
         r = 8;
         xs = Range[0., 1, 2^(-r)];
         (*Weights Dimension = 5 + 25 + 5, Biases Dimension = 5 + 5*)
         nExp = 45*100;
         activations = {Tanh, "Swish", "ReLU", "Sigmoid", Sin, "SELU"};
```

Generate simulations per activation function:

```
In[•]:=  ClearAll@generateNetwork;
         generateNetwork[activation_, seed_ : Automatic] :=
             NetInitialize[
                NetChain[
                   Append[Catenate@Table[{width, ElementwiseLayer@activation}, depth],
                      LinearLayer[{}, "Input" → width, "Biases" → 0.]], "Input" → "Real"],
                   Method → {"Random", "Weights" → 1., "Biases" → 1.}, RandomSeeding → seed];
```

```
In[•]:=  simulations = AssociationThread[ToString/@activations, ParallelTable[
               generateNetwork[act, baseSeed + i]@xs, {act, activations}, {i, nExp}]];
         Dimensions/@%
```

```
Out[•]=  <| Tanh → {4500, 257}, Swish → {4500, 257}, ReLU → {4500, 257},
            Sigmoid → {4500, 257}, Sin → {4500, 257}, SELU → {4500, 257} |>
```

```
In[•]:=  ClearSystemCache[];
         Share[];
```

Graphic options:

```
In[•]:=  SetOptions[FindDistribution, PerformanceGoal → "Quality"];
         SetOptions[ListPlot,
             PlotTheme → {"Detailed", "NeonColor"},
             PlotMarkers → {Automatic, Tiny}, PlotRange → All, PlotHighlighting → None
           ];
         SetOptions[ListLogPlot,
             PlotTheme → {"Detailed", "NeonColor"}, PlotRange → All, PlotHighlighting → None
```

```
    ];
SetOptions[ListLinePlot,
    PlotTheme → {"Detailed", "NeonColor"}, PlotRange → All, PlotHighlighting → None
    ];
SetOptions[Plot,
    PlotTheme → {"Detailed", "NeonColor"}, PlotRange → All, PlotHighlighting → None
    ];
customMatrixPlot[
    matrix_, labels_,
    OptionsPattern[{ImageSize → Medium, ColorFunction → "TemperatureMap"}]
    ] := MatrixPlot[
    matrix, PlotTheme → "Detailed", ColorFunction → OptionValue[ColorFunction],
    ImageSize → OptionValue[ImageSize], FrameTicks → {
        {Transpose[{Range[Length[labels]],
            Style[#, Bold, 14, "Subtitle"] & /@ labels}], None}, {None, Transpose[
            {Range[Length[labels]], Style[#, Bold, 14, "Subtitle"] & /@ labels}]}}
        }];
```

Activation Functions of Interest

We consider the following continuous nonlinearities:

```
In[ ]:=  actKeys = ToString /@ activations;
        actFns = AssociationThread[actKeys, {
            Function[x, Tanh[x]],
            Function[x, x/(1 + e^{-x})],
            Function[x, { x    x ≥ 0
                          0    True }],
            Function[x, 1/(1 + e^{-5 x/2})],
            Function[x, Sin[x]],
            Function[x, If[x ≥ 0, 1.0507 * x, 1.758099340847161 * x * (e^x − 1)]]
            }];
        Multicolumn[KeyValueMap[
            Plot[#2 @ x, {x, −2., 2}, Filling → 0., PlotLegends → None, PlotLabel → #1] &, actFns
            ], 2, Spacings → 2]
```

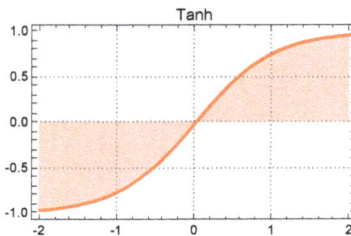

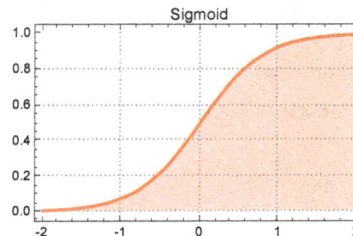

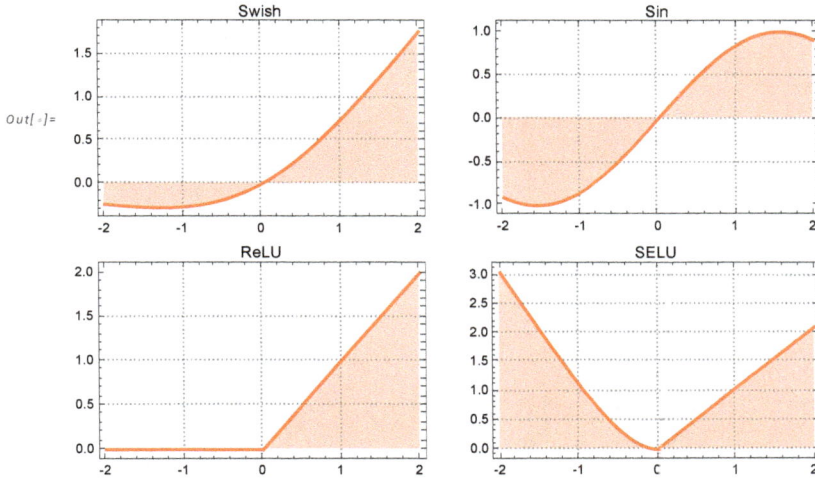

Differentiability

- Tanh, Swish, Sigmoid and Sin are infinitely differentiable

- Note that ReLU and SELU are not differentiable at 0, but infinitely differentiable elsewhere

First derivatives of the activations:

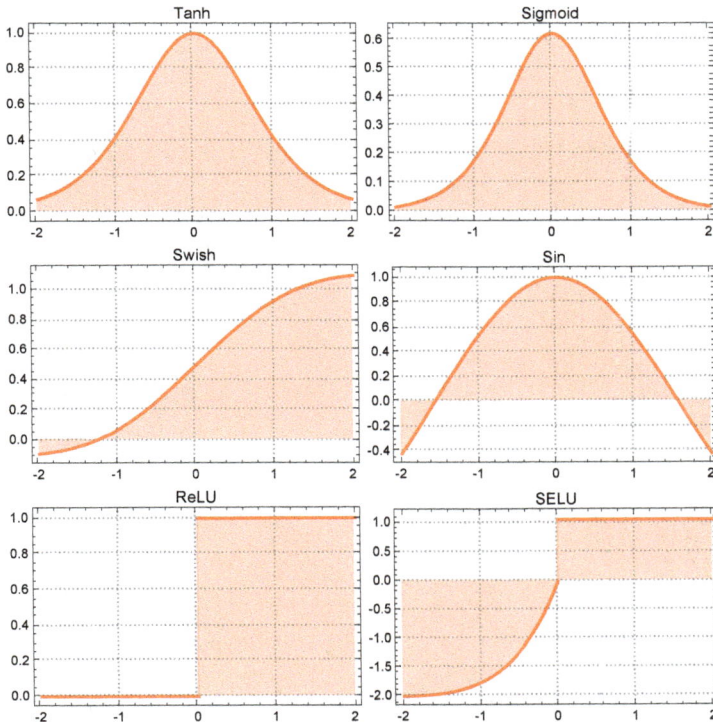

Second derivatives of the activations:

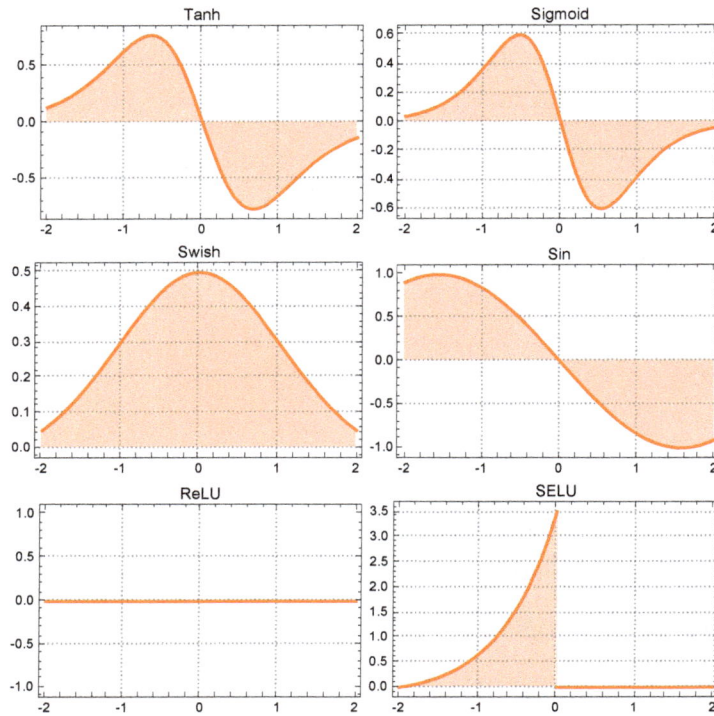

Fourier–Legendre

See [3] for Legendre polynomials and [4] for the Fourier–Legendre basis expansion.

The smooth activations (Tanh, Swish, Sigmoid and Sin) are well approximated on [–1, 1] by their Fourier–Legendre series of order 5:

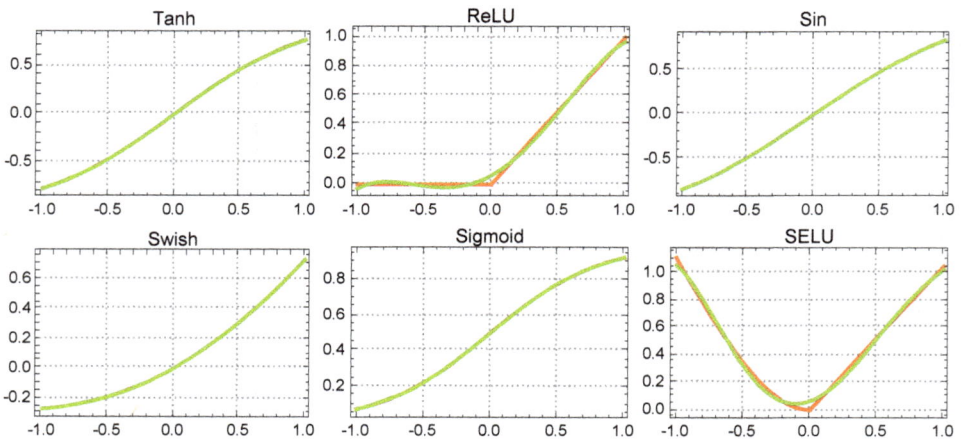

But when we increase the domain to [−2, 2], it requires expansions of order 9 to get good approximations for Tanh and Sigmoid. At order 9, we get also decent approximations for the singular ReLU and SELU:

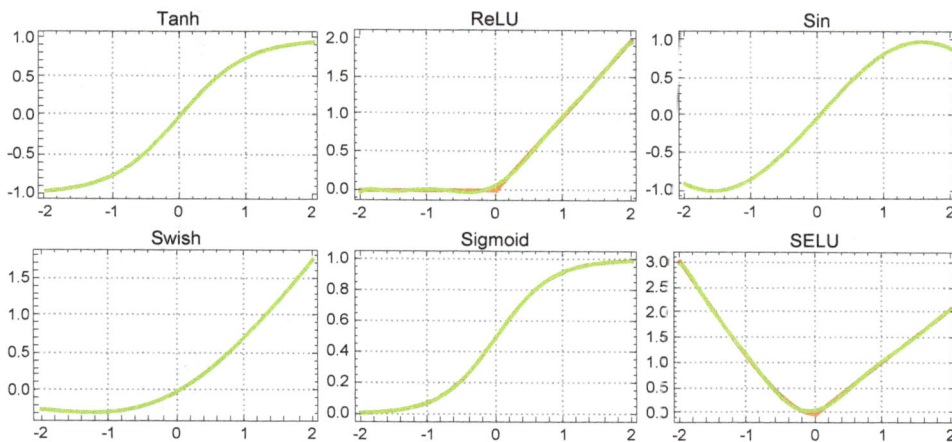

Discrete Wavelet Transform

See [5], [6] and [7] to review the discrete wavelet transform.

The following wavelet scalograms highlight the transform coefficients as rows of colorized rectangles, in which large absolute values are shown darker and each subsequent row corresponds to different wavelet index specifications:

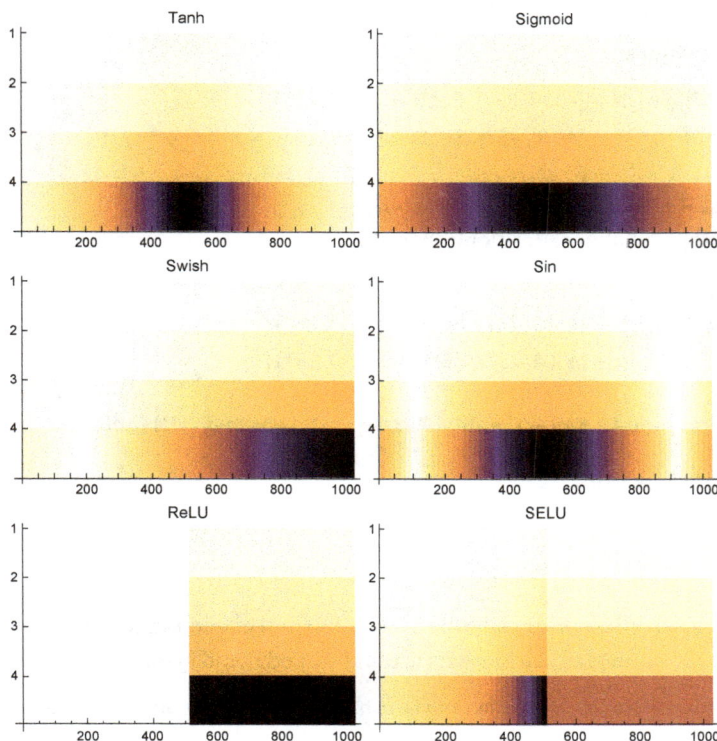

MLPs on [0, 1] with Varying Activation Functions

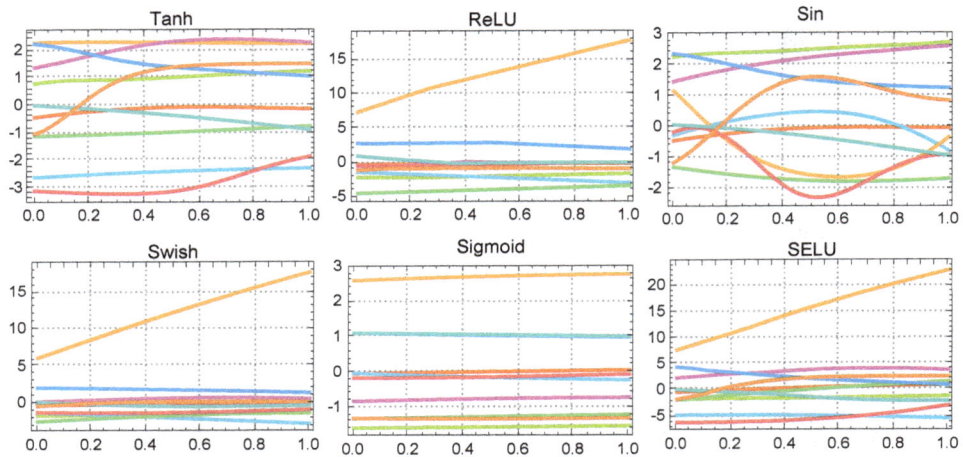

Exploratory Data Analysis

Preliminaries

We delete the constantly zero functions for ReLU:

In[•]:= **Dimensions /@ simulations**

Out[•]= **<| Tanh → {4500, 257}, Swish → {4500, 257}, ReLU → {4500, 257},**
Sigmoid → {4500, 257}, Sin → {4500, 257}, SELU → {4500, 257} |>

In[•]:= **range = Map[Max @ # – Min @ # &, simulations, {2}];**
simulations = Delete[simulations, Position[range, 0.]];
Dimensions /@ simulations

Out[•]= **<| Tanh → {4500, 257}, Swish → {4500, 257}, ReLU → {4414, 257},**
Sigmoid → {4500, 257}, Sin → {4500, 257}, SELU → {4500, 257} |>

In[•]:= **GroupBy[Position[range, 0.] /. Key → Identity, First → Last, Length @ # / N @ nExp &]**

Out[•]= **<| ReLU → 0.0191111 |>**

Distributions of Basic Statistics

We consider the following statistics: mean, standard deviation, norm and range (see the following). We observe that the mean distributions are all unimodal and symmetric:

- Swish, ReLU and SELU are more sharply peaked and have heavier tails than a normal distribution

- Tanh and Sigmoid are close to normal, with Sigmoid being slightly more concentrated
- Sin follows a logistic distribution, showing moderately heavy tails compared to normal

Moreover, the *scales* distributions are right skewed, and can be well approximated by a mixture of two gamma distributions or a mixture of one gamma distribution and one log-normal distribution.

Mean:

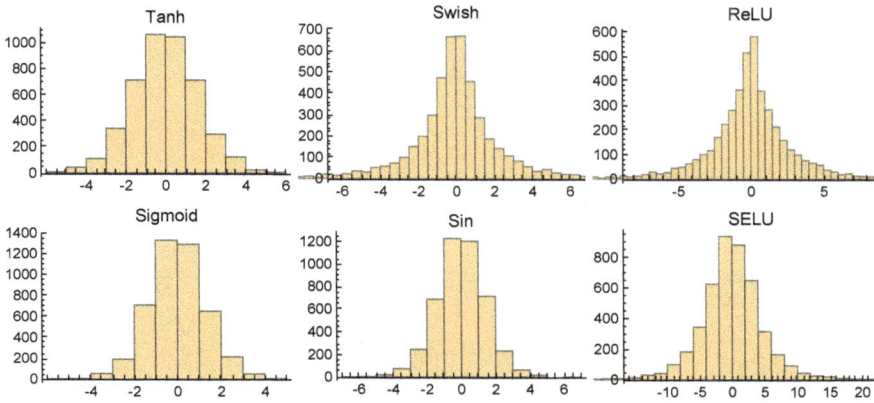

Tanh	NormalDistribution[-0.0272362, 1.64233]
Swish	StudentTDistribution[0.0136461, 1.21173, 1.87454]
ReLU	LaplaceDistribution[-0.0139919, 2.10476]
Sigmoid	NormalDistribution[-0.0168727, 1.279]
Sin	LogisticDistribution[-0.0130739, 0.795877]
SELU	StudentTDistribution[0.0114549, 3.35089, 4.38601]

Standard deviation:

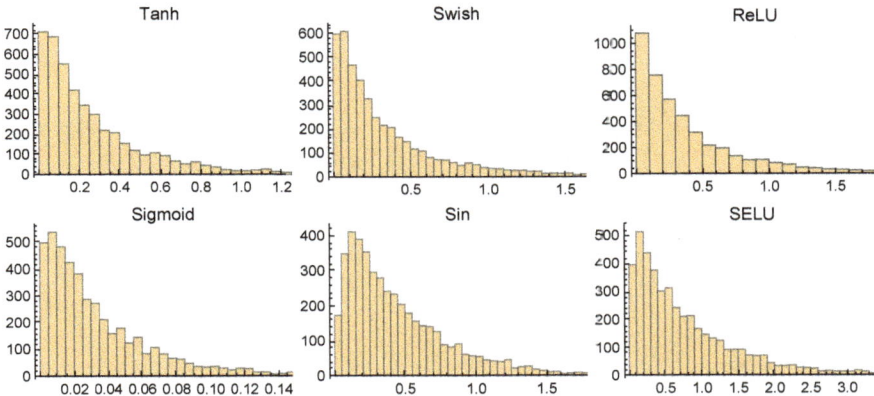

Tanh	ExponentialDistribution[3.47216]
Swish	MixtureDistribution[{0.894352, 0.105648}, {GammaDistr
ReLU	WeibullDistribution[0.839841, 0.429403]
Sigmoid	MixtureDistribution[{0.814354, 0.185646}, {BetaDistri
Sin	GammaDistribution[1.27121, 0.35669]
SELU	MixtureDistribution[{0.844456, 0.155544}, {GammaDistr

Norm:

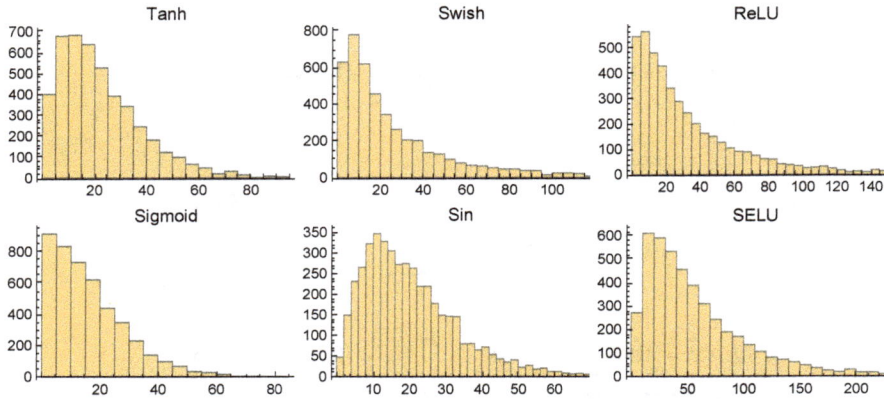

Tanh	GammaDistribution[1.85862, 12.6551]
Swish	MixtureDistribution[{0.853887, 0.146113}, {GammaDistr
ReLU	ExponentialDistribution[0.0283601]
Sigmoid	WeibullDistribution[1.25236, 17.3135]
Sin	GammaDistribution[2.2328, 9.42495]
SELU	MixtureDistribution[{0.868945, 0.131055}, {GammaDistr

Range:

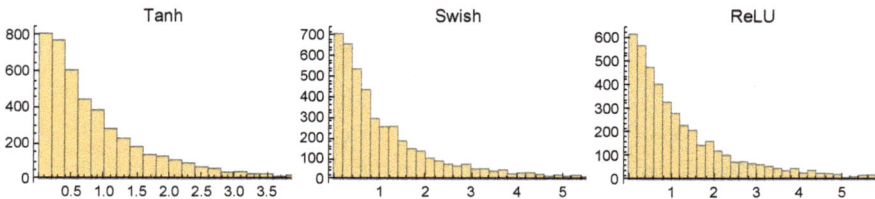

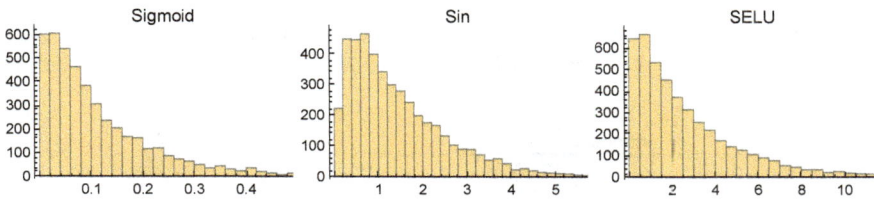

Tanh	ExponentialDistribution[1.06584]
Swish	MixtureDistribution[{0.907654, 0.0923459}, {GammaDist
ReLU	MixtureDistribution[{0.885021, 0.114979}, {Exponentia
Sigmoid	ExtremeValueDistribution[0.0636245, 0.0975458]
Sin	GammaDistribution[1.48721, 0.956553]
SELU	MixtureDistribution[{0.853139, 0.146861}, {GammaDistr

Empirical Estimates of the First Four Moments Functions

There is near-zero mean and skewness across all marginal distributions, except for SELU, which has a small skewness:

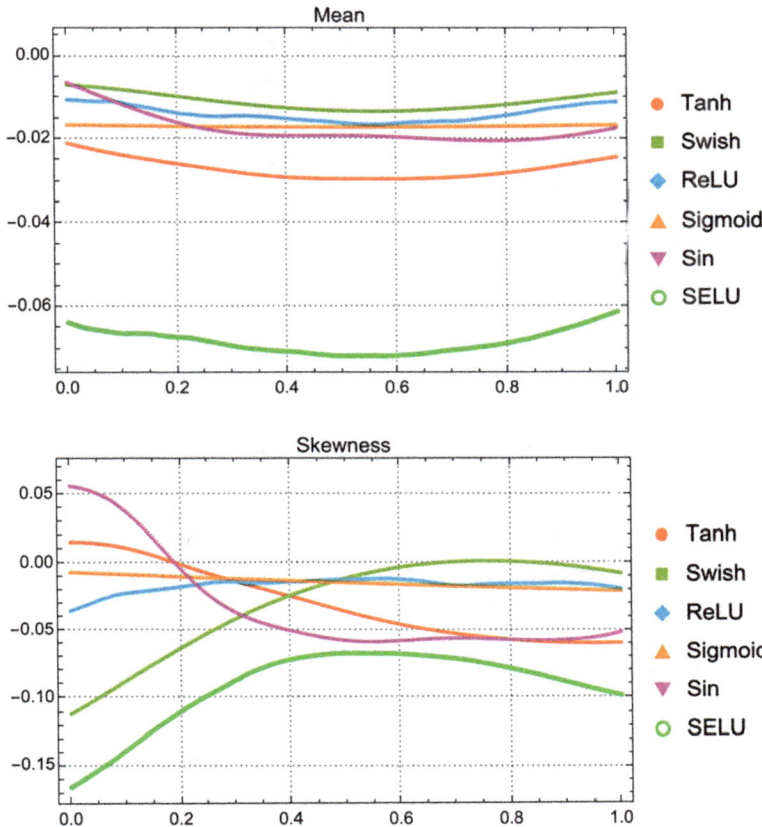

The variance and kurtosis clearly separate the activations Tanh, Sigmoid and Sin (for which they are rather constant with *small* values) from Swish, ReLU and SELU (for which they are increasing with *large* values):

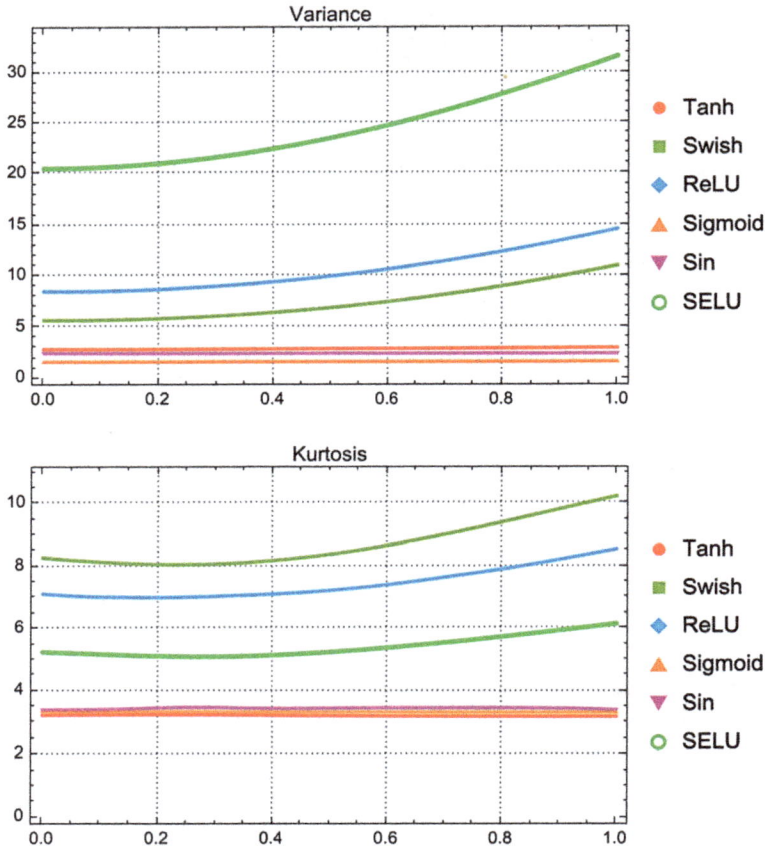

Karhunen–Loève Expansion

See [8] for a review of the Karhunen–Loève expansion.

It's an informed decision to perform the Karhunen–Loève expansion with a fixed number, in this case six, of principal functional components. Here's why.

We first selected the number of principal functional components based on the usual criteria of cumulated sums of eigenvalues on one hand and eigenvalues normalized by the first eigenvalue on the other. Then we had to deal with a lot of simulations not having a decent approximation through expansion, for which we looked at the principal functional components of their residuals. The results of the whole process were not far from retaining directly six principal functional components.

The principal functional components are smooth and well approximated by their Legendre basis expansion up to degree 9.

Principal Functional Components from Empirical Covariance Operator Estimate

Following are the first six eigenvalues of the empirical covariance operator:

Tanh	{0.938936, 0.0572547, 0.00320531, 0.000497369, 0.00008
Swish	{0.953934, 0.0448068, 0.00120835, 0.0000470009, 3.2506
ReLU	{0.962987, 0.0352078, 0.00146934, 0.000224525, 0.00006
Sigmoid	{0.998429, 0.00155993, 0.0000105822, 1.9927×10^{-7}, 4.22
Sin	{0.850589, 0.127831, 0.0183764, 0.00265783, 0.00047800
SELU	{0.945494, 0.053137, 0.00116344, 0.00014505, 0.0000362

And here are the corresponding principal functional components:

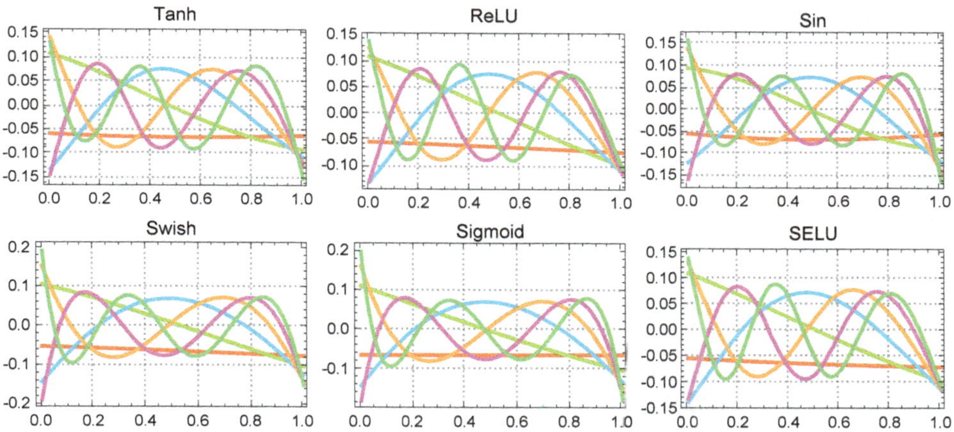

Similarity Measures of Generated Networks

The principal angles for the spaces are generated by principal functional components. See [7] and [8] for the the distance used:

	Tanh	Swish	ReLU	Sigmoid	Sin	SELU
Tanh	0	0.365352	0.154588	0.441192	0.266548	0.169311
Swish	0.365352	0	0.478656	0.161754	0.326996	0.447051
ReLU	0.154588	0.478656	0	0.575	0.400502	0.0988982
Sigmoid	0.441192	0.161754	0.575	0	0.298306	0.554521
Sin	0.266548	0.326996	0.400502	0.298306	0	0.412862
SELU	0.169311	0.447051	0.0988982	0.554521	0.412862	0

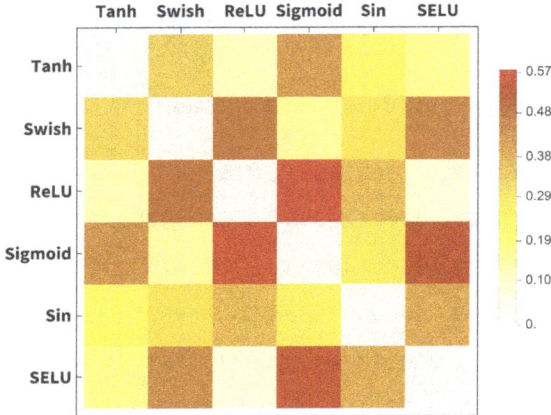

Here is the Legendre basis expansion of principal functional components (up to degree 9). See [3] for Legendre polynomials and [4] for the Legendre basis expansion. As we can see here, the errors are small:

Tanh	0.0271282
Swish	0.00187718
ReLU	0.0313206
Sigmoid	0.000443073
Sin	0.0032306
SELU	0.0255651

$$
\begin{pmatrix}
 & \text{Tanh} & \text{Swish} & \text{ReLU} & \text{Sigmoid} & \text{Sin} & \text{SELU} \\
\text{Tanh} & 0. & 0.107345 & 0.0673968 & 0.129558 & 0.116407 & 0.0525489 \\
\text{Swish} & 0.107345 & 0. & 0.120764 & 0.0473574 & 0.0968245 & 0.112771 \\
\text{ReLU} & 0.0673968 & 0.120764 & 0. & 0.144274 & 0.106814 & 0.0320879 \\
\text{Sigmoid} & 0.129558 & 0.0473574 & 0.144274 & 0. & 0.0872491 & 0.140356 \\
\text{Sin} & 0.116407 & 0.0968245 & 0.106814 & 0.0872491 & 0. & 0.118848 \\
\text{SELU} & 0.0525489 & 0.112771 & 0.0320879 & 0.140356 & 0.118848 & 0.
\end{pmatrix}
$$

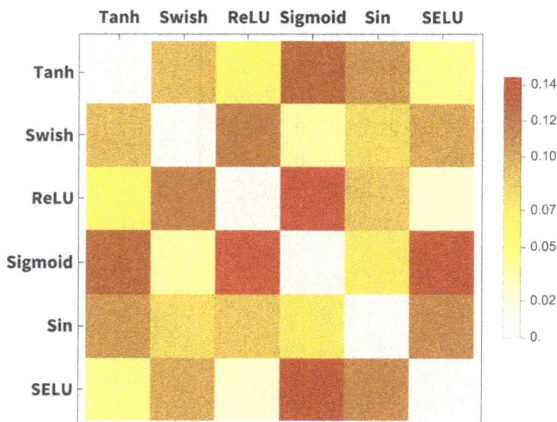

Following is a 2-nearest-neighbors graph:

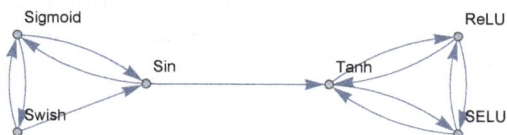

The proximity between spaces induced by the principal functional component coefficients is in accordance with the one from the distance matrix previously calculated. Moreover, we observe that:

- The space of networks generated by ReLU and SELU are close (closest pair), while the farthest pair is (ReLU, Sigmoid)
- The ones generated by Tanh and Sigmoid are also close, but to a lesser extent

Analysis of Approximation Truncated Expansion Errors

- As expected, there are large errors for ReLU and SELU given that they are not differentiable at zero. More precisely, at the 1% threshold, there is a proportion of errors of 30% for ReLU and 15% for SELU. We would need a very large expansion to approximate well the networks generated. In this case, the discrete wavelet transform is clearly an adapted framework to consider.

- For the other activations, the errors are quite small. The Swish and Sigmoid cases are well handled by the principal functional components (almost no errors at the 1% threshold). Outliers for Tanh and Sin provide interesting networks. To explore them, the Fourier–Legendre series expansion is clearly a more adapted framework to consider:

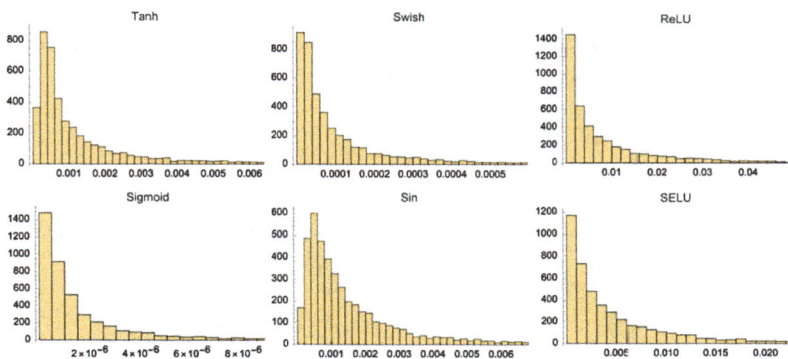

Following are the proportion of large relative errors (>1%):

Tanh	0.0364444
Swish	0.000888889
ReLU	0.300667
Sigmoid	0.0
Sin	0.0268889
SELU	0.153556

‹| Tanh → 164, Sin → 121 |›

In simulations using smooth activation functions, a number of networks exhibited abnormal behavior, defined as having relative errors greater than 1%. Specifically, 164 networks in the Tanh simulations and 121 networks in the Sin simulations were identified as abnormal. Examples of these networks are shown here:

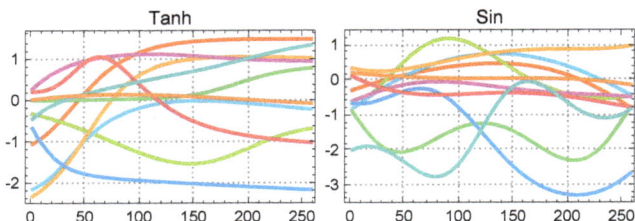

Learned Distribution for Smooth Activations Based on the Karhunen–Loève Expansion

- For the smooth activations (Tanh, Swish, Sigmoid and Sin), we learn approximated distributions as Gaussian mixtures for the random functions generated using their truncated Karhunen–Loève expansion coefficients

- The goodness of fit (as measured by cross-entropy) is much better for Sigmoid and Swish compared to Sin, with Tanh in between

- Surprisingly, the modes of the Gaussian mixtures don't correspond to *typical* networks (they exhibit very different shapes)

- *Atypical* networks (anomalies) are clearly more complex than *typical* networks and provide an interesting set for initialization to train a neural network on a complex task

Gaussian Mixture Estimates

Goodness of fit (cross-entropy):

$\langle |$ Tanh $\rightarrow 4.75 \pm 0.22$, Swish $\rightarrow 1.86 \pm 0.28$, Sigmoid $\rightarrow -15.47 \pm 0.11$, Sin $\rightarrow 8.60 \pm 0.10 |\rangle$

Following are the Gaussian modes for each process:

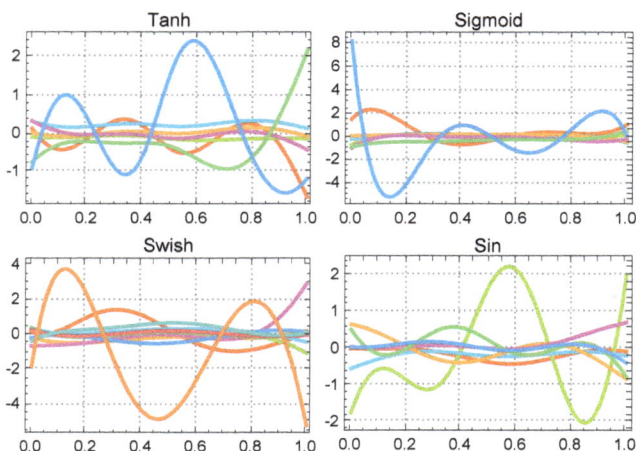

And here are some random variates from the learned distributions:

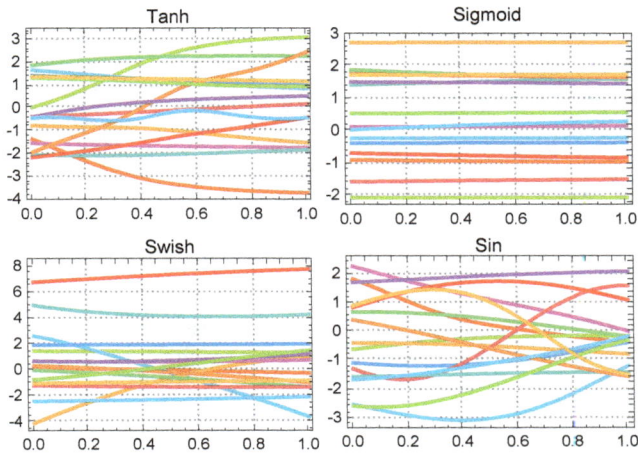

Core Dense Points: "Typical" Networks

Following are *typical* networks (high-density separated points):

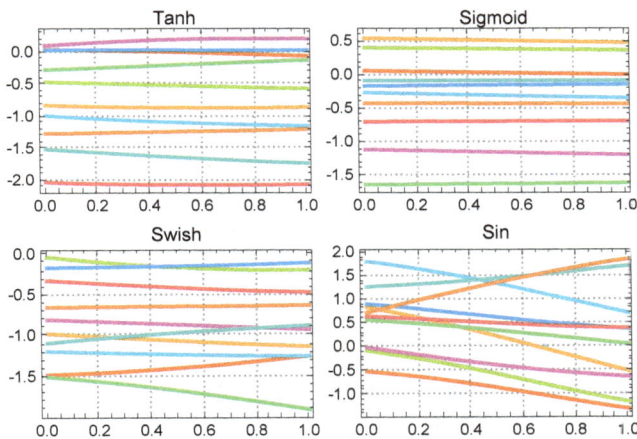

The graph of the typical networks shown previously clearly resembles the ones from the sampled MLPs in the section "MLPs on [0, 1] with Varying Activation Functions."

Anomalies: "Atypical" Networks

Following are the number of atypical nets in each simulation, out of a total of 4,500 networks per simulation:

Tanh	19
Swish	9
Sigmoid	16
Sin	13

And here are some of them:

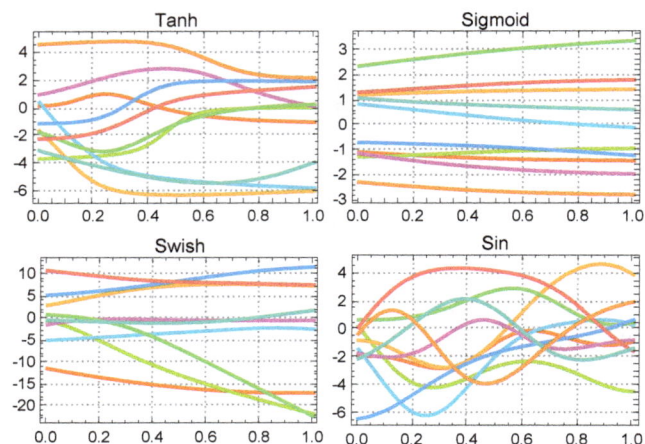

Atypical networks exhibit significant nonlinearity and variability in contrast to the typical ones.

Discrete Wavelet Transform to Explore the Spaces of Networks Generated by ReLU and SELU

See [5], [6] and [7] to review the discrete wavelet transform.

The computation of the discrete wavelet transform (Haar wavelet) of the networks at the maximum refinement level shows the following:

- The distributions of coarse coefficients are unimodal and symmetric, with a *thin* tail for ReLU (Laplace distribution) but *fat* for SELU (Student's t-distribution).

- The energy fractions of the detail coefficients are similar in terms of distributions. Their distributions are very peaked at the median.

- Almost all energy is concentrated at the last two details, the last one being largely dominant (with 19% and 74% median fractions, respectively).

- The projections (to a manifold of dimension 2) of the main wavelet coefficients have very similar shapes for ReLU and SELU.

- While the learned distributions for ReLU and SELU may appear different at first glance, a closer examination reveals they are more similar than they initially seem. The symmetric Kullback–Leibler divergence (Jeffreys divergence) between these two learned distributions is comparable to the divergence between the learned distribution for ReLU and its corresponding histogram distribution. This suggests that the differences fall within the range of empirical variation, indicating that the learned distributions for ReLU and SELU capture broadly similar underlying patterns.

- For SELU, the fit of the learned Gaussian mixture is good and makes sense given the shape of its projection (to a manifold of dimension 2).

Retrieve the Wavelet Transforms

```
In[•]:= dwds = ParallelMap[DiscreteWaveletTransform[Rest@#, HaarWavelet[], 8] &,
        KeySelect[simulations, MemberQ[{"ReLU", "SELU"}, #] &], {2}];
    ClearSystemCache[];
    Share[];
```

Coarse Coefficients

Distribution of the coarse coefficients:

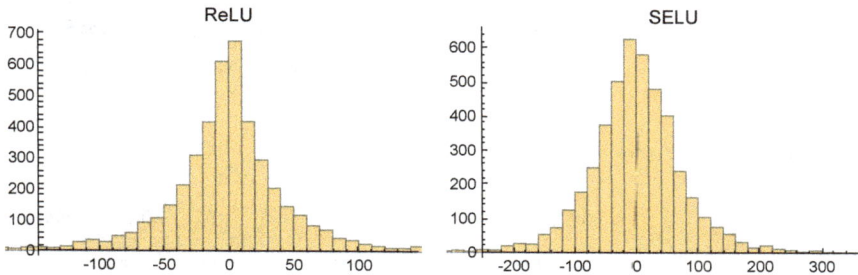

Skewness of the coarse coefficients:

ReLU	0.00307642
SELU	−0.0629516

Most likely distribution of the coarse coefficients:

ReLU	LaplaceDistribution[-0.224085, 33.6904]
SELU	StudentTDistribution[0.172138, 53.6536, 4.3903]

Energy Fraction of Details

Here are the details:

{1}
{0, 1}
{0, 0, 1}
{0, 0, 0, 1}
{0, 0, 0, 0, 1}
{0, 0, 0, 0, 0, 1}
{0, 0, 0, 0, 0, 0, 1}
{0, 0, 0, 0, 0, 0, 0, 1}

Mean of the energy fractions:

$\langle|$ ReLU →

{1}	0.000453758
{0, 1}	0.000787701
{0, 0, 1}	0.00246667
{0, 0, 0, 1}	0.00819822
{0, 0, 0, 0, 1}	0.0271185
{0, 0, 0, 0, 0, 1}	0.0854523
{0, 0, 0, 0, 0, 0, 1}	0.245994
{0, 0, 0, 0, 0, 0, 0, 1}	0.629529

, SELU →

{1}	0.0000744307
{0, 1}	0.000296154
{0, 0, 1}	0.00118066
{0, 0, 0, 1}	0.00468018
{0, 0, 0, 0, 1}	0.0180403
{0, 0, 0, 0, 0, 1}	0.0677302
{0, 0, 0, 0, 0, 0, 1}	0.231658
{0, 0, 0, 0, 0, 0, 0, 1}	0.67634

$|\rangle$

Median of the energy fractions:

$\langle|$ ReLU →

{1}	0.0000536177
{0, 1}	0.000214028
{0, 0, 1}	0.0008509
{0, 0, 0, 1}	0.00336689
{0, 0, 0, 0, 1}	0.0132356
{0, 0, 0, 0, 0, 1}	0.0508307
{0, 0, 0, 0, 0, 0, 1}	0.188517
{0, 0, 0, 0, 0, 0, 0, 1}	0.739583

, SELU →

{1}	0.0000494428
{0, 1}	0.000197597
{0, 0, 1}	0.00078892
{0, 0, 0, 1}	0.0031506
{0, 0, 0, 0, 1}	0.0125325
{0, 0, 0, 0, 0, 1}	0.0496786
{0, 0, 0, 0, 0, 0, 1}	0.196426
{0, 0, 0, 0, 0, 0, 0, 1}	0.736537

$|\rangle$

Distribution of the energy fractions per detail:

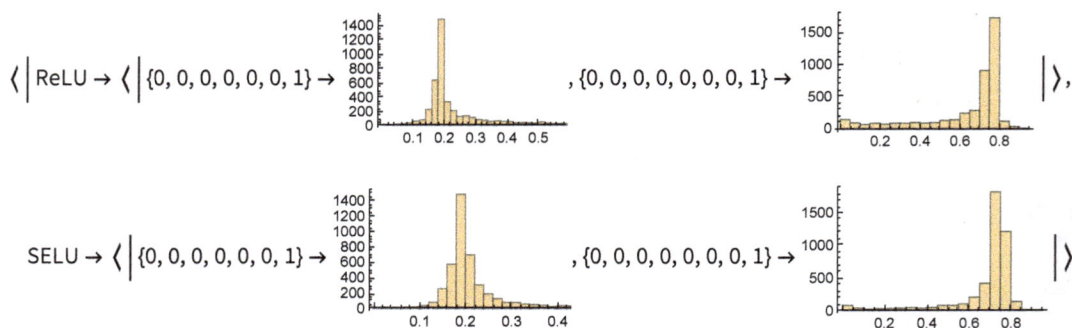

$\langle|$ ReLU → $\langle|$ {0, 0, 0, 0, 0, 0, 1} →

, {0, 0, 0, 0, 0, 0, 0, 1} →

$|\rangle$,

SELU → $\langle|$ {0, 0, 0, 0, 0, 0, 1} →

, {0, 0, 0, 0, 0, 0, 0, 1} →

$|\rangle$

Feature Space Plot of the Main Wavelet Coefficients

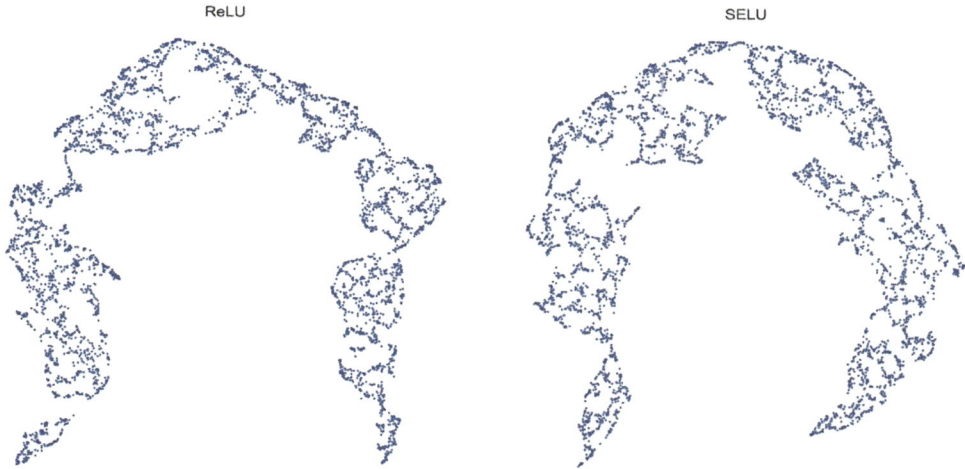

Learned Distribution: Gaussian Mixture Estimates

Goodness of fit (cross-entropy):

⟨| ReLU → 9.66 ± 0.14, SELU → 12.11 ± 0.11 |⟩

Kullback–Leibler divergence between the learned distribution and a corresponding histogram distribution for ReLU:

0.0462013

Kullback–Leibler divergence between the learned distribution and a corresponding histogram distribution for SELU:

0.00029252

Kullback–Leibler divergence between the learned distribution for ReLU and SELU:

0.0499274

Following are the Gaussian modes for each process:

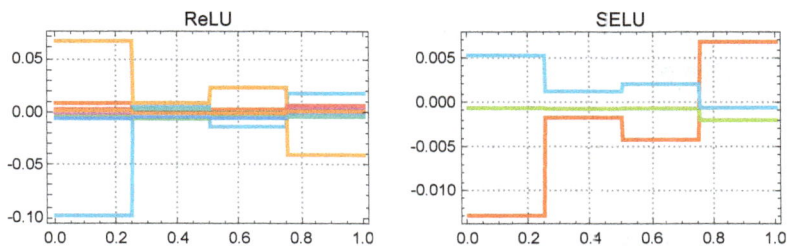

And here are some random variates from the learned distributions:

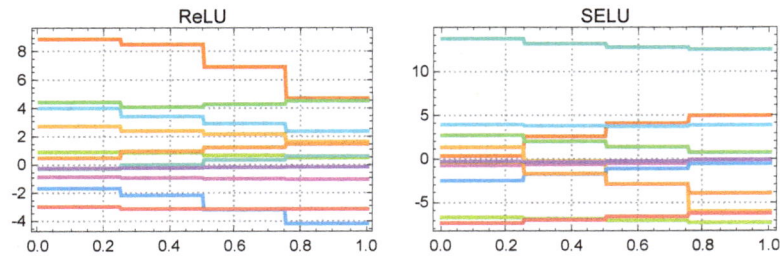

Concluding Remarks

This study has delved into the intricate world of shallow multi-layer perceptron networks, specifically focusing on the statistical and mathematical techniques to analyze, represent and compare distributions generated by classic activation functions. Here are the key takeaways from our observations and analyses:

- Distributions of basic statistics
  - All activation functions produced unimodal and symmetric mean distributions.
  - Swish, ReLU and SELU showed sharper peaks than normal, while Tanh closely followed a normal distribution.
  - ReLU, Sigmoid and Sin had thin tails, whereas Swish and SELU had fat tails.
  - The scale distributions were well approximated by a mixture of two gamma distributions or a mixture of one gamma and one log-normal distribution.

- Empirical estimates of the first four moments
  - Mean and skewness were minimal except for SELU, which showed small skewness.
  - The variance and kurtosis effectively distinguished the activations Tanh, Sigmoid and Sin (for which they are rather constant with *small* values) from Swish, ReLU and SELU (for which they are increasing with *large* values).

- Karhunen–Loève expansion
  - The Karhunen–Loève expansion was performed with six principal functional components based on cumulative eigenvalues and normalized eigenvalues. Despite initial simulation challenges, retaining six components provided a good balance between complexity and approximation accuracy.
  - The principal functional components were smooth and well approximated by their Legendre basis expansion up to degree 9.

- Similarity measures of generated networks
  - The proximity between spaces induced by the principal functional components' Legendre coefficients aligned with the distances measured using principal angles.
  - Networks generated by ReLU and SELU were the closest pair, while ReLU and Sigmoid were the farthest. Tanh and Sigmoid also showed some proximity, albeit to a lesser extent.

- Analysis of approximation truncated expansion errors
    - Large approximation errors were noted for ReLU and SELU due to their singularities at zero, with 30% and 15% of errors at the 1% threshold, respectively. This indicates the need for a large expansion, making the discrete wavelet transform a suitable alternative.
    - For smooth activations, errors were minimal. Swish and Sigmoid were well handled by the principal functional components. Outliers for Tanh and Sin presented interesting networks, better explored using the Fourier–Legendre series expansion.
- Learned distributions for smooth activations
    - For smooth activations (Tanh, Swish, Sigmoid, Sin), distributions approximated as Gaussian mixtures from truncated Karhunen–Loève expansion coefficients showed varying goodness of fit. Sigmoid and Swish had better fits than Sin, with Tanh in between.
    - Modes of the Gaussian mixtures did not correspond to typical networks; anomalies were more complex and potentially useful for initializing neural networks for complex tasks.
- Discrete wavelet transform for ReLU and SELU
    - The discrete wavelet transform revealed unimodal and symmetric distributions of coarse coefficients, with ReLU having thin tails and SELU having fat tails.
    - Energy fractions of detail coefficients were similarly distributed, with most energy concentrated in the last two details. The last detail was largely dominant.
    - Visualizations showed similar shapes for ReLU and SELU when projected onto a two-dimensional manifold. However, their learned distributions differed slightly, with the symmetric Kullback–Leibler divergence of the same magnitude as the one between the learned Gaussian mixtures and corresponding histogram distributions for ReLU.

In conclusion, this study provides a comprehensive analysis of the statistical properties of shallow multi-layer perceptron networks with various activation functions. The insights gained offer valuable guidance for initializing and understanding neural networks, particularly in terms of approximation errors, learned distributions and the effectiveness of different mathematical frameworks.

Acknowledgments

I would like to thank the staff of the Wolfram Summer School for providing this enriching opportunity and fostering such a stimulating environment. I am also thankful to Stephen Wolfram for his support of my project.

References

1. Wikipedia, "Universal approximation theorem." en.wikipedia.org/wiki/Universal _approximation_theorem.

2. N. J. Guliyev and V. E. Ismailov (2018), "Approximation Capability of Two Hidden Layer Feedforward Neural Networks with Fixed Weights," *Neurocomputing* 316: 262–269. doi.org/10.48550/arXiv.2101.09181.

3. E. W. Weisstein, "Legendre Polynomial," *Wolfram MathWorld*. mathworld.wolfram.com/LegendrePolynomial.html.

4. E. W. Weisstein, "Fourier–Legendre Series," *Wolfram MathWorld*. mathworld.wolfram.com/Fourier-LegendreSeries.html.

5. S. Mallat (1998), *A Wavelet Tour of Signal Processing*. Academic Press.

6. Wolfram Research (2017), DiscreteWaveletTransform, Wolfram Language function. reference.wolfram.com/language/ref/DiscreteWaveletTransform.html.

7. Wikipedia, "Haar wavelet." en.wikipedia.org/wiki/Haar_wavelet.

8. J. Wang, et al. (2015), "Review of Functional Data Analysis." arxiv.org/abs/1507.05135.

9. J. Van Dyke (2020), "Distances between Subspaces" [Presentation given at the University of Texas at Austin]. web.ma.utexas.edu/users/vandyke/notes/deep_learning_presentation/presentation.pdf.

10. K. Ye and L.-H. Lim (2016), "Schubert Varieties and Distances between Subspaces of Different Dimensions." *SIAM Journal on Matrix Analysis and Applications* 37(3). doi.org/10.48550/arXiv.1407.0900.

Access the Full Code

Scan or visit wolfr.am/WSS2024-Benarous.

Cite This Notebook

"Exploring the Distributions of Shallow Multi-layer Perceptrons"
by Teo Benarous
Wolfram Community, STAFF PICKS, July 9, 2024
community.wolfram.com/groups/-/m/t/3213030

The Sand Reckoner, Now in Wolfram Language

YAU SHAN YANG

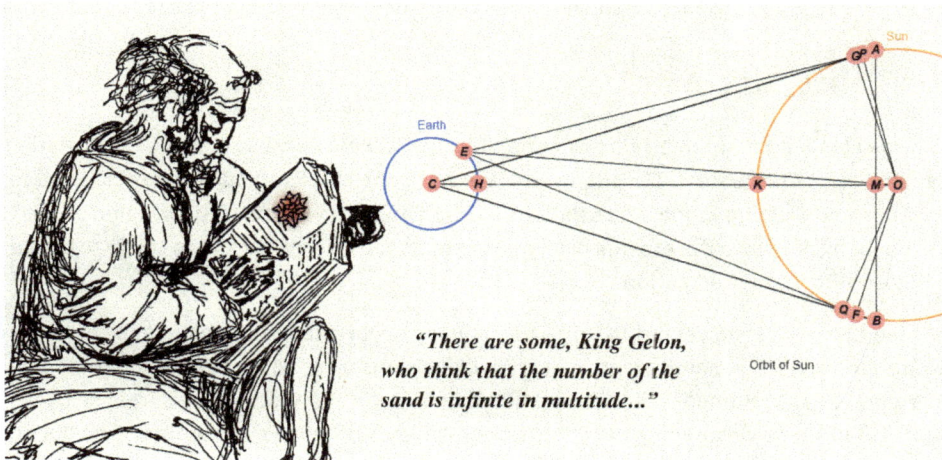

"There are some, King Gelon, who think that the number of the sand is infinite in multitude..."

This project aims to recreate Archimedes's Psammites (The Sand Reckoner) using Wolfram Language, making it more accessible to a wider audience by means of Archimedes's friendly tone and interactive diagrams. The goal is to imbue the reader with both a sense of satisfaction in understanding the text and a sense of appreciation of Archimedes's incredible work, all while assuming minimal mathematical knowledge. There is still some additional work to be done to complete a full representation of The Sand Reckoner, which will be pursued after the Wolfram Summer School.

Introduction

About *The Sand Reckoner*

Archimedes wrote *The Sand Reckoner* around the third century BC. The eight-page paper addressed to King Gelon II of Syracuse sets out to determine the upper bound of grains of sand that would fit inside the universe, with the end goal of demonstrating Archimedes's system to express extremely large numbers that remedies the issue of "innumeracy" of the Greek numerical notation system.

The work is interesting to the modern audience in many ways:

1. It is one of the first-known examples of psychophysics (the study of observing the capabilities of humans).

2. It contained many details about ancient astronomy.

3. The ancient Greeks, having no formal mathematical notation, used sentences to describe the relationships between quantities. It is remarkable how they managed to discover so much without modern mathematical symbols.

4. It is written for the layman, potentially one of the earliest examples of a research paper.

Recreating the Work

This project is not meant to be a translation of *The Sand Reckoner*, but rather a computational essay that aims to explain it. Despite its reputation as one of Archimedes's most accessible works, I found existing English translations a little difficult to follow. This was one of the early challenges I faced during this project. However, it did motivate me to really focus on the digestibility of my recreation.

I decided that my version should also be narrated by Archimedes, as such a tone can achieve a friendlier exposition. Some creative narrative decisions were taken that are not based on the original text. Archimedes didn't explain the Greek number system in his treatise. He would also never have known what chopsticks were, nor have ever eaten pizza. The decisions were made to make certain concepts easier to understand.

I did not manage to complete the essay during the Summer School, with current progress standing roughly at 50% of the original content. Following is what has been done so far.

The Sand Reckoner

Introduction

How much sand do you think there is on Earth? Some people think it is infinite, and others say that it is innumerate—which is to say there is so much sand that we don't have a number that can describe it.

Well, I think both these types of people are wrong, and I can tell you why.

Whoops! I have forgotten to introduce myself. *Χαῖρε*! My name is Archimedes, or *Ἀρχιμήδης* if you write it in Greek. If I was alive today (in 2024), I would be celebrating my $2,311^{\text{th}}$ birthday.

Number Systems

I think the reason why people during my time thought that sand is innumerable is because our Greek counting system is rather clunky. Your current system only uses 10 symbols (0, 1, 2, 3, 4, 5, 6, 7, 8 and 9). You use each one to represent the quantities zero to nine and then use the position of the symbols to represent whether you mean ones, tens, hundreds, thousands, etc.

Your number system, also known as Western Arabic numerals, is a fantastic way to represent numbers. You only have to remember 10 symbols. Our ancient Greek number system is not as simple. We need about 28 symbols and a few special markings. Here is a table of the symbols:

1	2	3	4	5	6	7	8	9
"α"	"β"	"γ"	"δ"	"ϵ"	"$\digamma$"	"ζ"	"η"	"θ"
10	20	30	40	50	60	70	80	90
"ι"	"κ"	"λ"	"μ"	"ν"	"ξ"	"o"	"π"	"ϱ"
100	200	300	400	500	600	700	800	900
"ρ"	"σ"	"τ"	"υ"	"ϕ"	"χ"	"ψ"	"ω"	"$\jmath$"
1000	2000	3000	4000	5000	6000	7000	8000	9000
"$_{\iota}\alpha$"	"$_{\iota}\beta$"	"$_{\iota}\gamma$"	"$_{\iota}\delta$"	"$_{\iota}\epsilon$"	"$_{\iota}\digamma$"	"$_{\iota}\zeta$"	"$_{\iota}\eta$"	"$_{\iota}\theta$"

Our number system is additive, which is to say you add the symbols together to get your number. "M" means 10,000; "$_{\iota}\theta$" means 9,000; "σ" means 200; and "α" means 1—so 19,201 is written as "$M_{\iota}\theta\sigma\alpha$" in Greek. You might have noticed that we use the same symbols ($\alpha, \beta, \gamma, ...$) for our alphabet as well. So to distinguish between the two, we would add a bar above the number like this: "$M_{\iota}\bar{\theta}\sigma\alpha$".

You also might have noticed that the table does not tell you how to write numbers above 19,999. How do we express numbers like 20,000, 30,000 and so on? Well, we Greeks regard 10,000 (or a *myriad*) as the unit for counting the big numbers. To represent 20,000, we will write $\overset{\bar{\beta}}{M}$. The β above the M represents 2, so $\overset{\bar{\beta}}{M}$ means $2 \times 10{,}000$, or 20,000. This way, we can get quite large numbers, including those that are multiples of 10,000, like $9 \times 10{,}000$ ($\overset{\bar{\theta}}{M}$), $50 \times 10{,}000$ ($\overset{\bar{\nu}}{M}$) and $200 \times 10{,}000$ ($\overset{\bar{\sigma}}{M}$), all the way up to $10{,}000 \times 10{,}000$ ($\overset{\bar{M}}{M}$). You can now try out different numbers and see what they look like in Greek:

Type a number! 173 678

$$\overset{\overline{\iota\zeta}}{M}_{\iota}\gamma\chi o\eta$$

What next? How are we Greeks going to count to numbers more that ten thousand ten thousands? Well, if they so wish, they can use a system that I have invented to talk about large numbers—I mean *huge* numbers. Numbers that, if written in your number system, will have a 1 followed by at least a quadrillion zeros (note that we Greeks back then didn't have the word "quadrillion").

But before we get to my fancy number system, which I have already told my friend Zeuxippus (or *Ζεύξιππος*) all about, I want to address the problem of counting sand. As I said before, some people think that there is so much sand on Earth that there is no way we can use numbers to describe it, but I will show you that my number system can describe not only the number of grains of sand that can fill the Earth but also the number of grains of sand that can fill the *entire universe*.

Don't believe me? Well, let's get to it.

Estimating the Size of the Universe

Firstly, how big is the entire universe? Well, during our time, we aren't all too sure. The Greek astronomer Aristarchus (*Ἀρίσταρχος*) thought that the Sun was at the center of the universe, with the Earth orbiting around it and the furthest point of the universe being a sphere inlaid with the stars:

```
In[ ]:= Framed[Manipulate[Module[{earthTexture, earthRadius, earthLabelPosition,
            moonColour, moonRadius, moonLabelPosition, sunTexture,
            sunRadius, sunLabelPosition, starSphereTexture, kosmos},
          earthTexture = Texture[          ];

          earthRadius = 12;
          earthLabelPosition = {170, 50, 50};
          moonColour = Gray;
          moonRadius = 8;
          moonLabelPosition = {130, 30, -30};

          sunTexture = Texture[          ];

          sunRadius = 39;
          sunLabelPosition = {50, 50, 50};

          starSphereTexture = Texture[          ];

          kosmos = Graphics3D[{
                {sunTexture, Sphere[{0, 0, 0}, sunRadius],
                  Style[Text["Sun", sunLabelPosition], FontSize -> 16, Orange, Bold],
                  Style[Line[{sunLabelPosition - 2, {0, 0, 0}}], Thick, Orange]},
                {earthTexture, Sphere[{120, 0, 0}, earthRadius],
                  Style[Text["Earth", earthLabelPosition], FontSize -> 16, Blue, Bold],
                  Line[{earthLabelPosition - 2, {120, 0, 0}}]},
                {moonColour, Sphere[{160, 0, 0}, moonRadius],
```

```
        Style[Text["Moon", moonLabelPosition], FontSize → 16, Black, Bold],
        Style[Line[{moonLabelPosition+6, {160, 0, 0}}], Thick, Black]},
      {starSphereTexture, Sphere[{0, 0, 0}, 800],
        Style[Text["The Sphere of Stars i.e the Edge of the Universe", {0, 0, 900}],
          FontSize → 16, Purple, Bold]}}, ViewPoint → Front,
    ImageMargins → −80, Background → Lighter[LightBrown, 0.8]];


    Show[kosmos, PlotRange → zoomout*50, Boxed → False, ImageSize → Large]],
  {{zoomout, 1, Style["Zoom out!", Medium]}, 1, 20},
  ControlPlacement → Top,
  FrameLabel → Style["Aristarchus's Universe", Large, Bold], Paneled → False,
  SaveDefinitions → True], Background → Lighter[LightBrown, 0.8]]
```

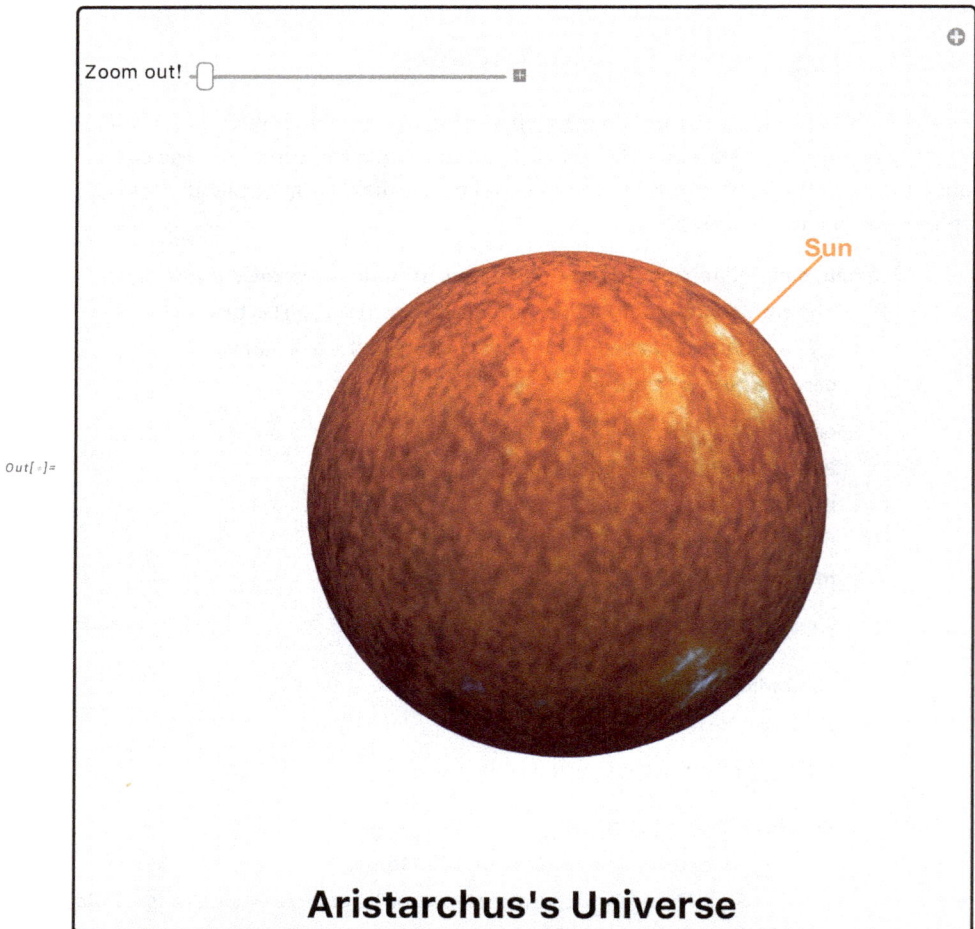

Out[·]=

I'm going to borrow that idea of the universe and modify it a little. During my time, what we considered as the "universe" is the world in which the Earth is at the center with the Sun orbiting around it, and the furthest point of this universe is the very orbit of the Sun:

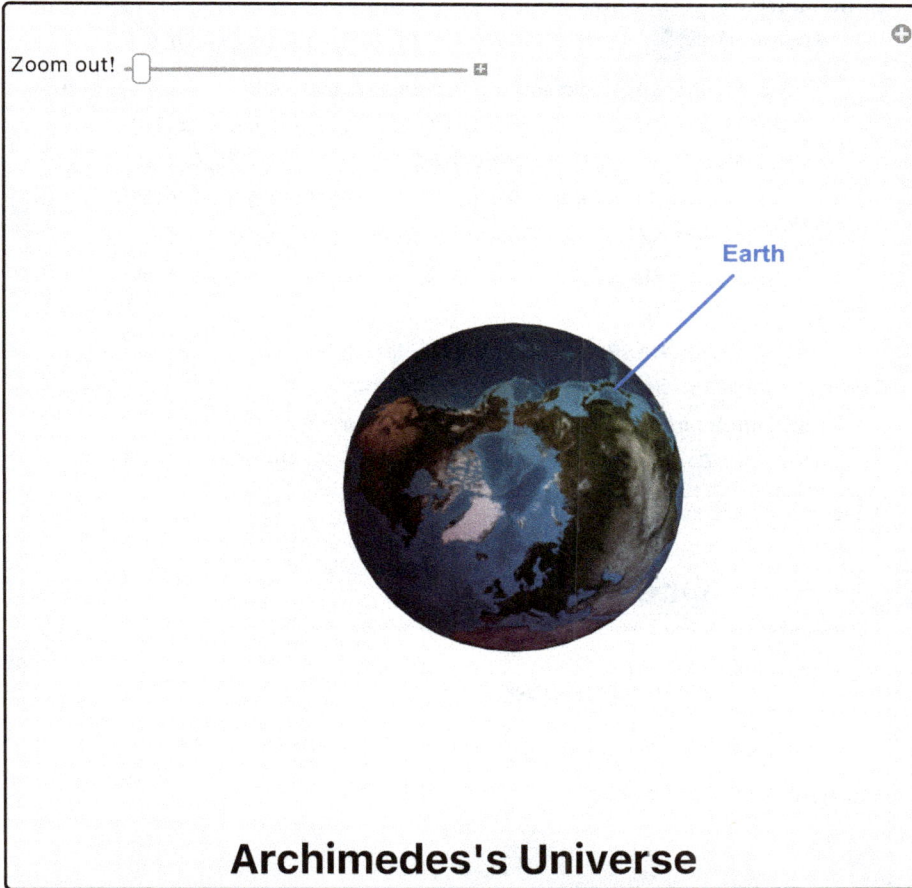

Archimedes's Universe

To estimate the size of this model of the universe, I am going to assume four things.

First, I'm going to say that the diameter of the Earth is three hundred myriad stadia (3,000,000 in your numbers, $\bar{M}$ in Greek numbers). *Stadia* is the unit we Greeks use to calculate distances, just like kilometers and miles, and one stadium is around 180 meters:

```
In[ ]:= stadiaMetresDiagram = Graphics[
          {
            Style[
              Text["One stadium", {180/2, 8}], Bold, Large],
            Line[{{0, -2}, {0, 0}, {180/2, 0}, {180/2, 3}, {180/2, 0}, {180, 0}, {180, -2}}],
            {Brown, AbsoluteThickness[4], Line[{{0, -5}, {180, -5}}]},
            Style[Text["180 metres", {180/2, -26}], Bold, Large], Line[{{0, -18}, {0, -20},
                {180/2, -20}, {180/2, -23}, {180/2, -20}, {180, -20}, {180, -18}}],
            {Orange, AbsoluteThickness[4], Dashing[1/180], Line[{{0, -15}, {180, -15}}]}
          },
          ImageSize -> Medium];
```

```
earthDiameterDiagram = Graphics3D[
     {Texture[        ], Rotate[Sphere[{0, 0, 0}, 180], −100 Degree, {0, 0, 1}],
        {Brown, AbsoluteThickness[10], Dashing[1/3 000 000],
           Rotate[Line[{{−180, 0, 190}, {180, 0, 190}}], 90 Degree, {0, 1, 0}]},
        {AbsoluteThickness[2], Rotate[Line[{{−180, 0, 150}, {−180, 0, 211},
              {0, 0, 211}, {0, 0, 220}, {0, 0, 211}, {180, 0, 211}, {180, 0, 150}}],
           90 Degree, {0, 1, 0}]}, {Texture[300 myriad stadia],
        Rotate[Cuboid[{−180, 0, 220}, {180, 4, 260}], 90 Degree, {0, 1, 0}]}
     },
     Boxed → True, ViewPoint → Front, ImageSize → Medium];
Column[{"Stadia vs. metres\n", stadiaMetresDiagram,
     "\n\nDiameter of the Earth", earthDiameterDiagram},
   Frame → True, Background → Lighter[LightBrown, 0.8], ItemSize → Small]
```

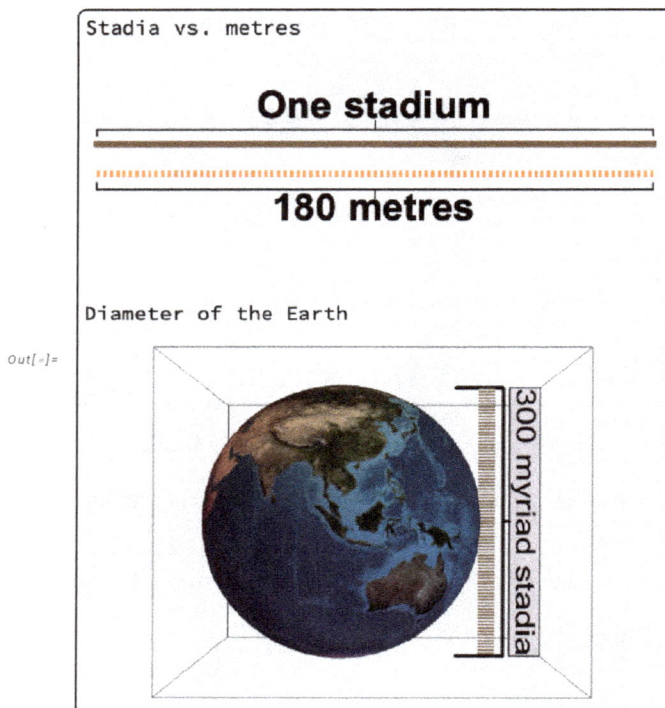

This three hundred myriad stadia is actually about 10 times more than the estimate for Earth's diameter agreed upon among Greek astronomers, but I am not trying to be accurate here. I want to make sure that the universe that I am describing is as big as possible, so that it can contain the most grains of sand possible, so that I can show off a huge number using my number system.

Second, I'm going to say that the Sun is bigger than the Earth, and that the Earth is bigger than the Moon. This is something I'm sure we won't argue too much about.

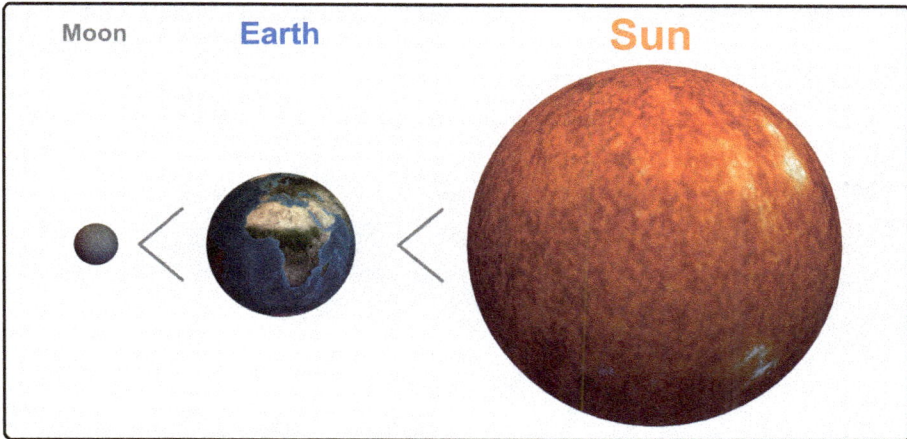

Third, I will say that the Sun's diameter is 30 times the Moon's diameter. Some Greek astronomers before me said that it is nine times larger. My father Phidias ($\Phi\varepsilon\iota\delta\iota\alpha\varsigma$) said that it is 12 times larger. I, playing it safe and large, will say it is 30 times larger.

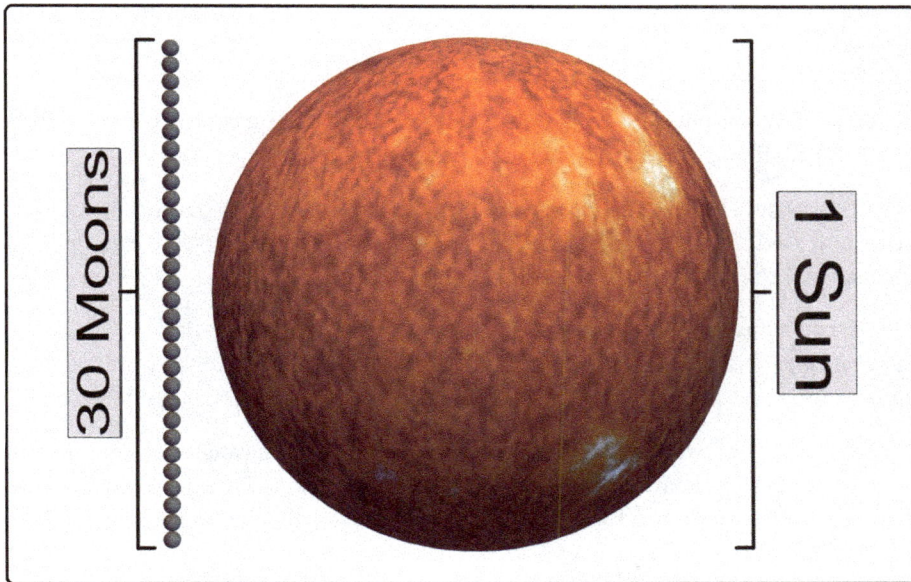

Fourth, I am going to say that the Sun's diameter is larger than a side of a 1,000-sided polygon (also known as a chiliagon) that is inscribed (fits inside) its orbit. You may play with the following slider to see what I mean:

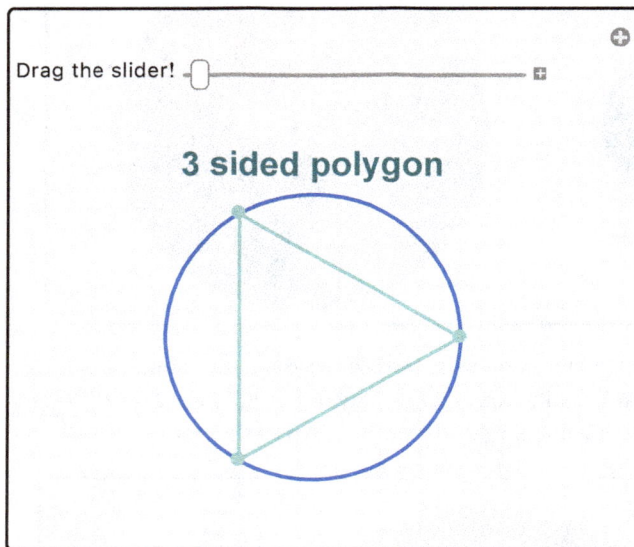

Drag the slider!

3 sided polygon

This last assumption is not something I just made up. It is based on careful observations and geometric calculations, which I will show you now.

I had to first measure the *angular diameter* of the Sun. If you had the longest chopsticks in the world and used them to "clamp" the sides of the Sun while standing on Earth, the angle made between those chopsticks is the angular diameter.

I would share how exactly I measured the angular diameter of the Sun, but it will take a while, so perhaps some other time! The main thing to know now is that the result of my measurements is that the angular diameter of the Sun is between 0.45 and 0.54 degrees.

Now we are going to draw a diagram of the Earth and Sun. C is the center of the Earth and O is the center of the Sun. Remember that we are using my version of the universe, so the Sun orbits the Earth. Take note that the image is also not to scale, and that the Sun should be waaaaay farther away from the Earth:

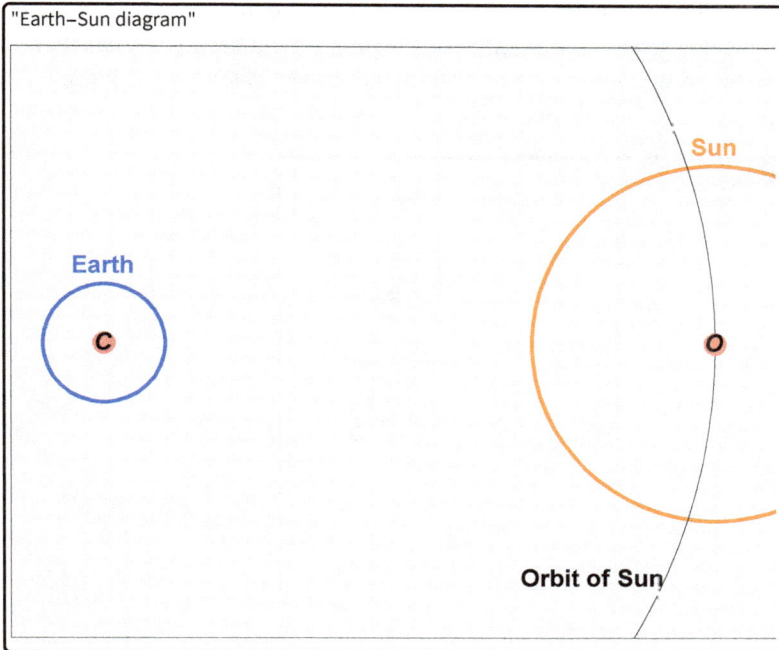

"Earth–Sun diagram"

Remember the part about chopsticks and measuring the angular diameter of the Sun? We are going to draw a point E on the Earth to represent where such a chopstick-wielder would be standing and draw two lines representing the chopsticks from E to the sides of the Sun, P and Q. The angle between the chopsticks, which we can also call angle PEQ, will be the angular diameter of the Sun. To repeat, angle PEQ, according to my super-accurate measurements, is between 0.45 and 0.54 degrees:

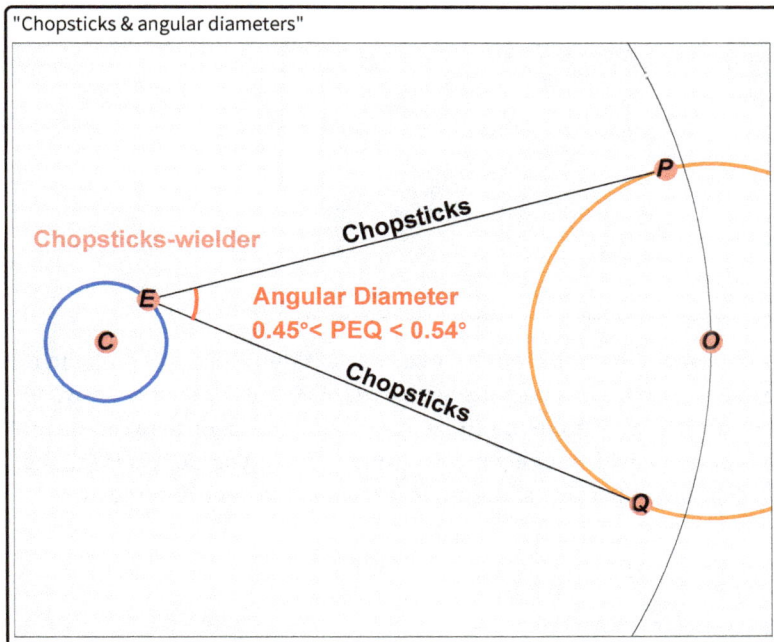

"Chopsticks & angular diameters"

Now imagine that the chopsticks-wielder is standing in the center of the Earth. The angular diameter of the Sun will have changed slightly. Do you think it will be larger or smaller than angle PEQ? You can test your ideas out using some actual chopsticks, along with something round:

Finished with your chopsticks investigation? Another way to think about this problem is to first see that the distance E to O is less than the distance C to O, and the farther away the point where the chopsticks meet is from the circle, the smaller the angle between them. You could continue to play around with your chopsticks. Alternatively, you can play with the following slider. It also shows the value of the angle formed at the top-left corner:

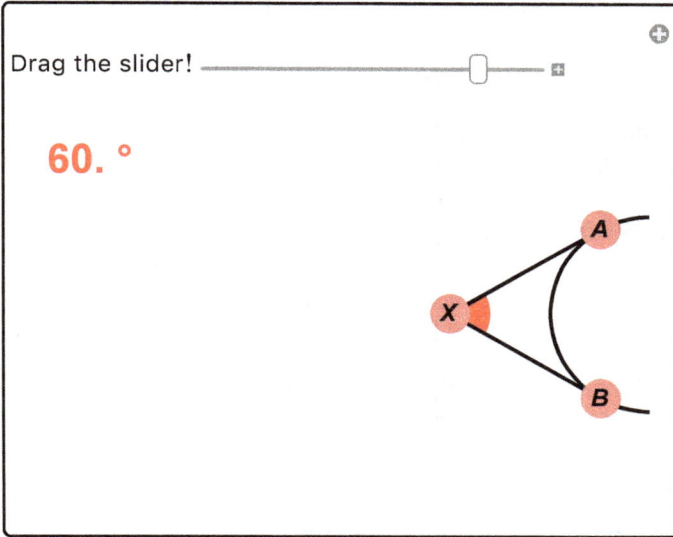

Bringing together all we know from these investigations, we can then say that the angle GFC is smaller than PEQ. And since angle PEQ is smaller than 0.55 degrees, angle GFC also has to be smaller than 0.54 degrees. In summary:

$0.45° < PEQ < 0.54°$
$GCF < PEQ$
$GCF < 0.55°$

Now let's extend the imaginary chopsticks that start from point C. Extend them all the way till they meet the orbit of the Sun, which isn't extending too much. It is important to note that A and B are different from G and F; they look similar because they are very close, but G and F lie on the circle of the Sun while A and B lie on the thin line that is the Sun's orbit. We are also going to draw a line between A and B:

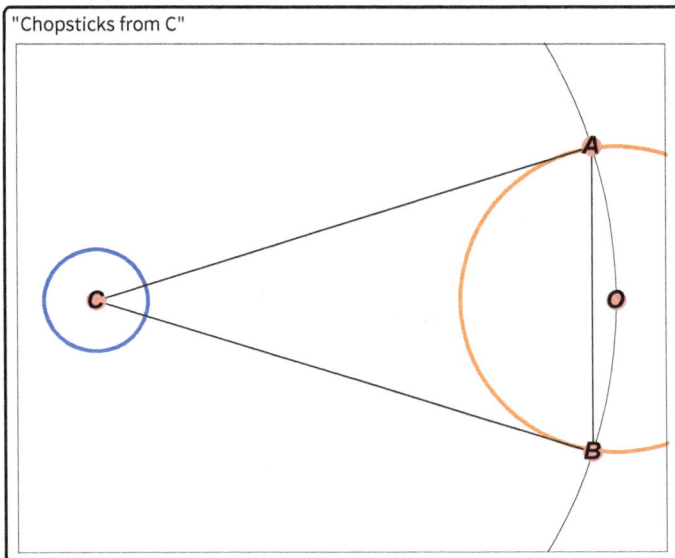
"Chopsticks from C"

The angle formed by the two chopsticks should now be exactly the same as angle GCF, because we haven't done anything to the chopsticks except for extending their ends. This angle, angle ACB, should also be less than 0.45 degrees.

I suppose that you know that turning your head all the way around is 360 degrees, whereas turning it halfway is 180 degrees. Two 180 degrees can fit inside 360 degrees because $180 + 180 = 360$. That's why you need two half-turns to make a full turn. If your head turned 60 degrees instead, you would have to turn six times, as there are six 60s that fit into 360, as $60 + 60 + 60 + 60 + 60 + 60$ (or 6×60) is 360. Now here is a question for you: how many times would you have to turn your head 0.55 degrees in order for you to turn all the way around (360 degrees)? It is a rather big number (bonus points if you can write it down in Ancient Greek!).

If you had said about 654 or 655, you would be correct! You need about 654-ish 0.55 degrees to make 360. Using this thinking and turning back to our diagram, if we had many copies of the triangle ACB and put them in the circle like slices in a pizza, how many triangles would fit in the circle? You can use the following sliders to help you with the problem:

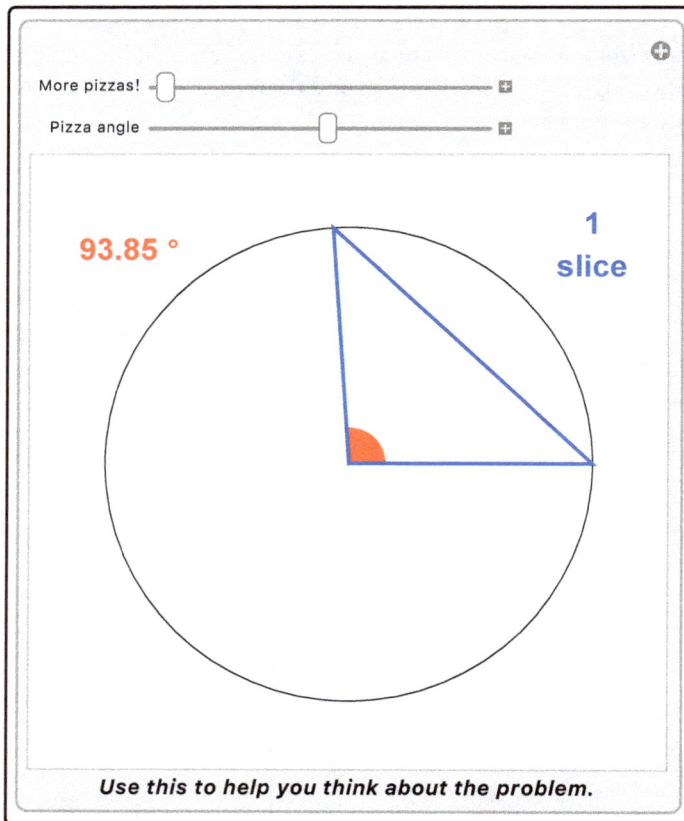

Use this to help you think about the problem.

Got it? The answer will also be about 654 triangles. After you fill up the circle with pizza triangles, they also sort of form a many-sided shape. If the pizza angle is 90 degrees, since you can fit four 90 degrees in 360 degrees, you get four triangle pizza slices in the circle, forming a sort of square. With 0.55 degrees, you can fit about 654 triangles inside the circle, forming a 654-ish-sided shape.

Going back to our diagram of the Earth and Sun, and taking what we learned about pizza triangles, since the pizza triangle ACB in our diagram has a pizza angle smaller than 0.55 degrees, a pizza triangle with side AB should be smaller than the side of the 654-ish-sided shape that fits in the circular orbit of the Sun:

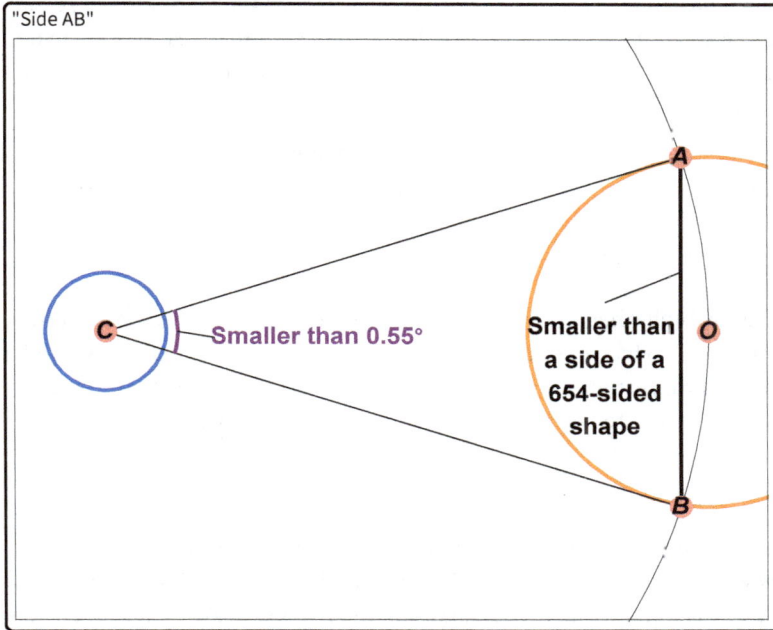

"Side AB"

Smaller than 0.55°

Smaller than a side of a 654-sided shape

You may have realized that the perimeter of this 654-ish-sided shape (or any shape that is fit inside of a circle) is kind of similar to the perimeter of the circle (also known as the circumference). The more sides the shape has, the more similar it will be to the circle.

Here is an interesting fact about circles: if you cut a bit of string with the length of a circle's radius (which is the distance from the center of a circle to its sides), it will take around six and little more (about $6\frac{1}{7}$) of these string pieces to wrap around the circle's circumference.

… *to be continued.*

Conclusion

That Which Remains to Be Done

I would divide the entire text of *The Sand Reckoner* into the following parts:

1. Opening (about innumeracy)

2. Assumptions 1, 2, 3 and "assumption" 4

3. Measuring the angular diameter of the Sun

4. Geometric proof of "assumption" 4

5. Calculation of the size of the universe and amount of sand

6. Archimedes's number system

Of these six sections, I have done sections 1, 2 and half of section 4. I have skipped section 3 as it felt deserving of a separate essay of its own. The rest are all things that remain to be completed. Further refinement with regards to the smoothness of animations, clarity of language, etc. are also places for this project to improve.

It is also possible that I may have misunderstood or grossly misrepresented certain parts of *The Sand Reckoner*. Please do let me know if I have made any mistakes in this regard.

That Which Could be Done

There are a couple extra things that one might consider adding to make the essay even richer—for instance, a function with which one can input one's own values for the initial assumptions and that outputs the resulting estimate for the upper bound of grains of sand that would fit the assumed universe. LLM integration could also be a potential avenue for exploration.

Closing Thoughts

I will be honest and say that at the start of this project, I thought that I would be done in about a week. How wrong I was! The process of understanding Archimedes, learning how to code in Wolfram Language, fixing the myriad and seemingly inexplicable bugs that came up in the code… all these things took time! That said, I had loads of fun solving the problems that came up and regret nothing.

A fun thought is to imagine if Archimedes had access to Mathematica during the third century BC. What would he have conjured up? At the very least, he could rejoice in the fact that no one will disturb his circles that he had drawn up in Graphics.

Appendix

Initialization Cells and Extra Code

Code That Gives the Coordinates Where a Tangent Meets a Circle Given a Point on the Tangent, and Radius and Position of the Circle

```
In[•]:=  tanCoord[{x_, y_}, {a_, b_}, r_] :=
            Module[{dToCenter, tanLength, simulSol, T, J, answerCoords},
               dToCenter = EuclideanDistance[{x, y}, {a, b}];
               tanLength = Sqrt[dToCenter ^ 2 - r ^ 2];
               simulSol = Solve[(T - a) ^ 2 + (J - b) ^ 2 == r ^ 2 &&
                     Sqrt[(x - T) ^ 2 + (y - J) ^ 2] == tanLength, {T, J}, Reals];
               Thread[{T /. simulSol, J /. simulSol}]]
```

Notes: This function occasionally gives warnings from the Solve function. It probably has something to do with the square roots and stuff, but I haven't figured it out.

Visualization of a Line Wrapping a Circle

```
In[ ]:= Manipulate[Graphics[{Thick, Circle[{0, 0}, 10], Orange, Circle[{0, 0}, 10, {-Pi/2, a - Pi/2}],
        Rotate[Style[Line[{{0, -10}, {10 (2 Pi - a), -10}}], VertexColors →
            {Red, Orange, Yellow, Green, Blue, Purple}], Thick, Orange], a, {0, 0}]},
      PlotRange → {{-10 Pi, 20 Pi}, {-11, 15 Pi}}], {{a, 0, "test"}, 0, 2 Pi, 0.001}]
```

Out[ ]=

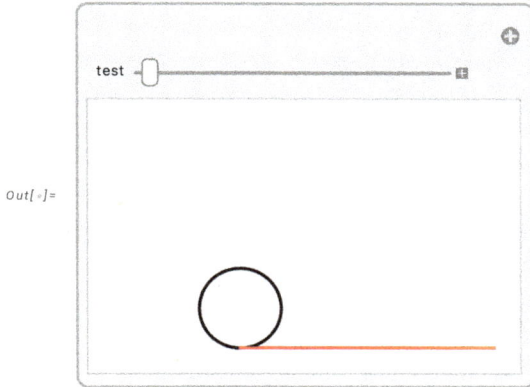

Notes: My plan is to use this Manipulate to demonstrate the ratio between the radius and circumference of a circle. I wish to add an animation to this Manipulate where the radius of the circle "flies out" and forms the orange line whose length is the circumference of the circle.

Complete Geometric Construction for the Proof of Assumption 4

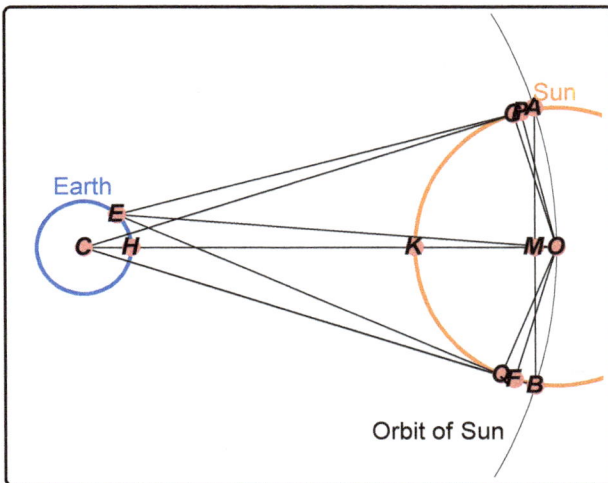

Notes: This is the main template I used to make the geometric diagrams to prove Archimedes's fourth assumption in *The Sand Reckoner*.

Bloopers

Experiments on the "Pizza" Manipulate

During the process of bug fixing for the "pizza" visualization, I encountered some weird and wonderful behaviors of the Manipulate. I have not given too much thought as to why they behave the way they do, but they look cool, so here they are, raw and unaltered:

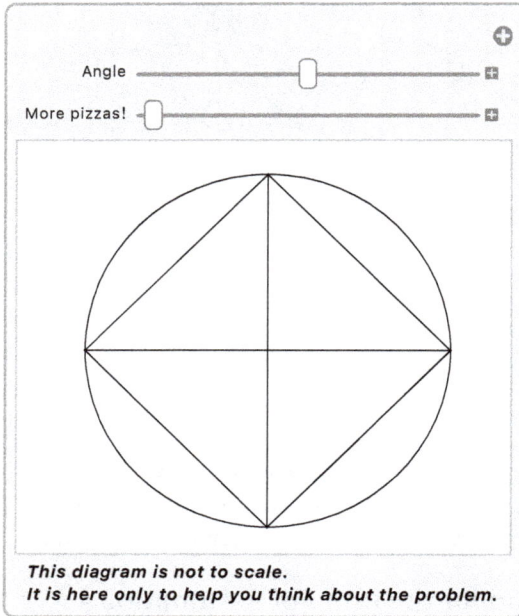

This diagram is not to scale.
It is here only to help you think about the problem.

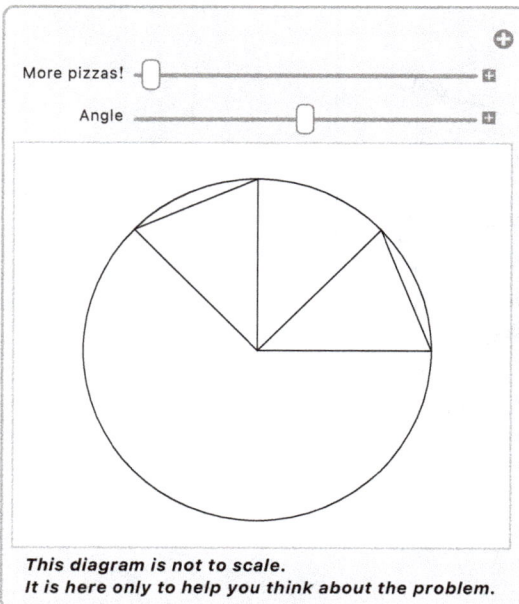

This diagram is not to scale.
It is here only to help you think about the problem.

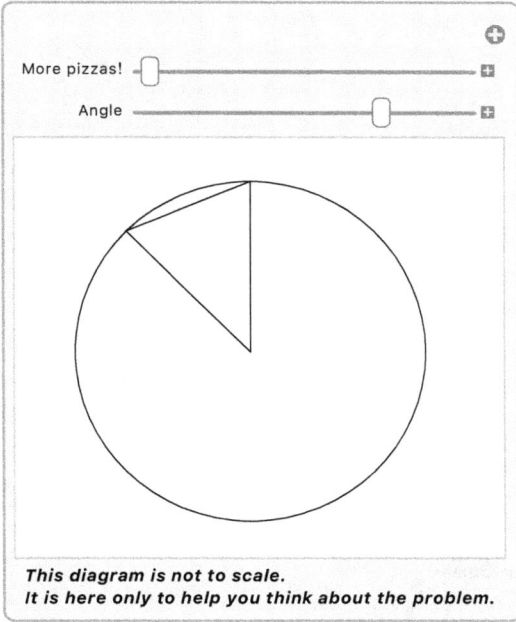

This diagram is not to scale.
It is here only to help you think about the problem.

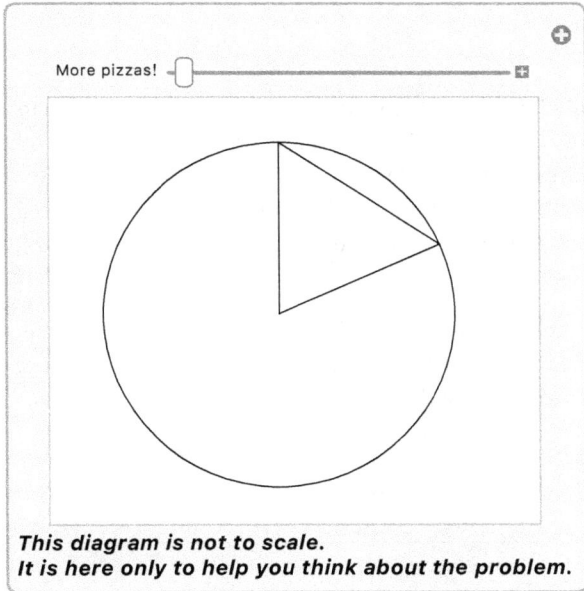

This diagram is not to scale.
It is here only to help you think about the problem.

Angle

More pizzas!

Here is a diagram to help you think about the problem.

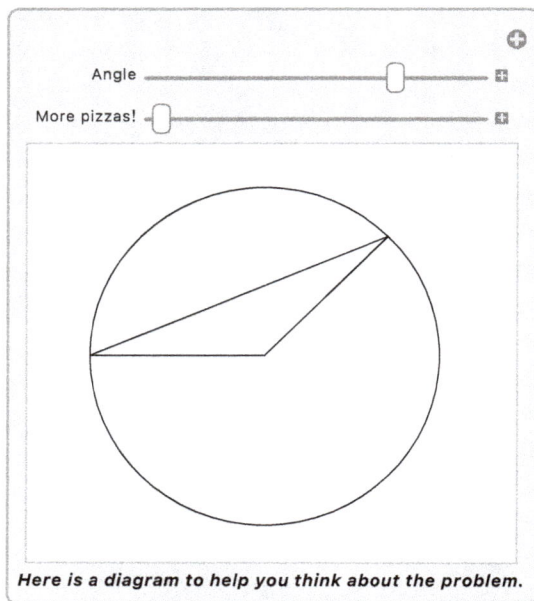

Acknowledgements

There are just so many people I would like to thank. I would like to thank my mentors John McNally and Faizon Zaman, as well as Phileas Dazeley-Gaist, Bob Nachbar and all the other mentors and coursemates in the Wolfram Summer School for patiently guiding me through my project and helping me out when I had problems or difficulties. I would also like to express gratitude to Stephen Wolfram and Christopher Wolfram for suggesting such an interesting project to me.

A super-duper special and sincere thanks to the Shah family, without whom none of this would be possible. Thank you, Mr. Shah, for your incredible support, and thank you too, Margot and Milly, for being my motivation to continue learning and sharing what I love.

Last but not least, I would like to say thank you to my family, for all the love, care and support, which I feel and appreciate deeply in my heart.

References

1. Wikipedia, "Greek numerals. en.wikipedia.org/wiki/Greek_numerals.

2. G. J. Toomer (2024), "Archimedes," *Encyclopedia Britannica*. www.britannica.com/biography/Archimedes.

3. Archimedes, *Ψαμμίτης* (in Greek). (Translated in T. Heath (1897), as *The Sand Reckoner*, Cambridge University Press. sacred-texts.com/cla/archim/sand/sandreck.htm.)

4. O. Thomas (2022), "Archimedes' Psammites ('Sand-Reckoner') Part 2: How Much Sand Would Fill the Universe?" *YouTube*. www.youtube.com/watch?v=_NApkn7nxmI.

5. I. Vardi, *Archimedes, the Sand Reckoner*, Internet Archive. archive.org/details/archimedessandreckonervardi.

6. H. Mendell, "Archimedes, Sand-Reckoner (Arenarius), Introduction to the Translation," California State University, Los Angeles. web.calstatela.edu/faculty/hmendel/Ancient%20Mathematics/Archimedes/SandReckoner/SandReckoner.html.

7. Wikipedia, "*The Sand Reckoner*." en.wikipedia.org/wiki/The_Sand_Reckoner.

8. B. Carroll, "The Sand Reckoner," Weber University. physics.weber.edu/carroll/Archimedes/sand.htm.

Access the Full Code

Scan or visit https://wolfr.am/WSS2024-Yau.

Cite This Notebook

"*The Sand Reckoner*, Now in Wolfram Language"
by Yau Shan Yang
Wolfram Community, STAFF PICKS, July 9, 2024
community.wolfram.com/groups/-/m/t/3210729

Analyzing Multiway Chemical Reaction Networks Using Chemical Space Networks

NICHOLAS FRIELER

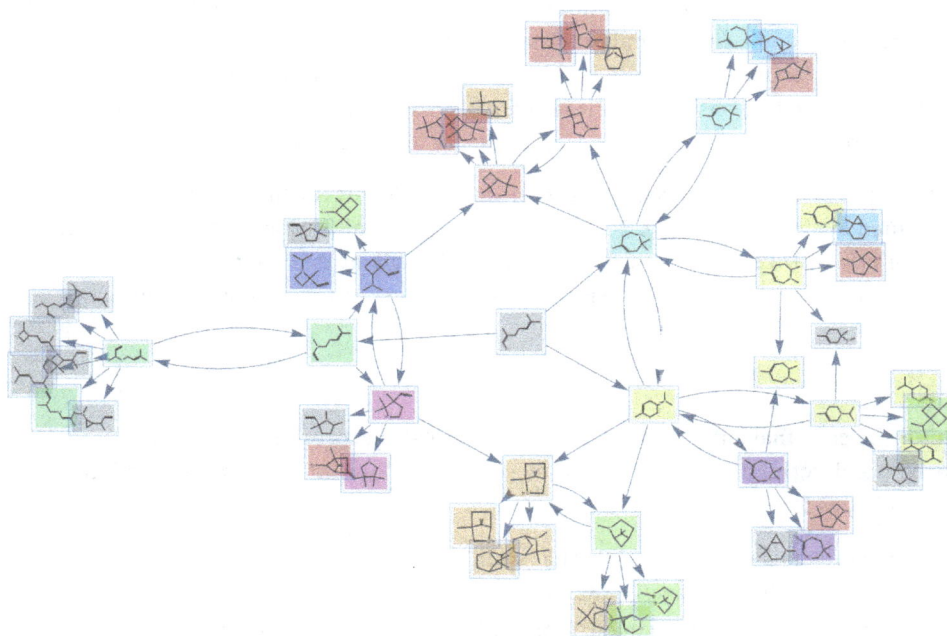

This project combines multiway chemical reaction networks with chemical space networks to analyze structural similarities between the products of different reaction pathways with identical or closely related precursor molecules. It explores several different avenues for analysis that can be paired with these methods, including the number of clusters in chemical space networks, the size of the largest such cluster and branchial graph analysis. It also verifies the validity of chemical space network clustering using multiway genealogical distance and principal component analysis on the molecules produced by the multiway chemical reaction network.

Introduction

Visualization is one of the most important tools when it comes to data analysis. It allows researchers to see clear trends in the data they are working with, and different visualization techniques allow for different features of the data to be analyzed. In the past decade, chemical space networks have emerged as a powerful visualization tool for chemists and biologists. A chemical space network is a collection of molecules, each represented as a node, with an edge between two molecules if they are sufficiently similar. Chemical space networks are very versatile, as they can be used on systems with a very large number of molecules, and they can be used with any similarity metric, allowing for different features of molecules to be analyzed [1].

In the world of computation, exhaustive searching is an essential tool for fully exploring a given system. In Wolfram Language, this can be achieved with multiway graphs, which take in a list of states and iteratively apply a set of rules for a specified number of steps. The generality of multiway computation allows it to be applied to a wide array of scenarios. Previously, Ariana-Dalia Vlad completed a project developing a multiway model of chemical reactions. This allows for a great number of related molecules to be generated based on only a few reactions and a small number of starting molecules [2]. In a multiway chemical system, all molecules present will be structurally related in some way, with the level of similarity depending on the reactions used. This makes the molecules from any multiway chemical reaction network a perfect candidate for analyzing using chemical space networks.

This project is heavily inspired by the work done by Robert Nachbar, especially what he presented in his 2022 Wolfram Technology Conference talk, "Modeling Chemical Reaction Networks with Token-Event Graphs" [3]. In fact, his talk presented multiway analysis on monoterpene biosynthesis, which is the main chemical system explored in this post.

MoleculePlotOutputForm

This is the code that automatically displays Molecule objects using MoleculePlot rather than the typical descriptor box. See the output generated below this cell for information on how to disable this temporarily:

```
In[ ]:=  moleculeImageSize[mol_Molecule, scale_ : 0.5] :=
            Module[{atoms, bonds, coords, dist, minmax},
                atoms = mol["AtomList"] /. Atom[a_, ___] :> a //
                    AssociationThread[Range@Length@#, #] &;
                bonds = Select[mol["BondList"] /. Bond[ij_, ___] :> ij,
                    FreeQ[atoms /@ #, "H"] &];
                coords = QuantityMagnitude@mol["AtomDiagramCoordinates"];
                dist = EuclideanDistance @@ coords[[#]] & /@ bonds //
                    Replace[{} -> {0}] // Mean;

            minmax = coords[[#]] & /@ bonds // Catenate // Map[First] // MinMax;

            Max[36, Min[6 × 72, (Subtract @@ Reverse@minmax)/Max[1, dist] scale 72]]
                ]

        $RBNMoleculeFormat = True;

        Print["Set $RBNMoleculeFormat = False to restore built-in behavior."];
```

```
Chemistry`Formatting`moleculeBox[mol_, fmt_] /; TrueQ[$RBNMoleculeFormat] :=
     If[Chemistry`Common`has3DCoordinates[mol],
          With[{graphics = MoleculePlot[mol]},
               {plot = ToBoxes[graphics, fmt]},

          Replace[plot, {HoldPattern[Graphics3DBox[x_, y___]] :↦
                         Graphics3DBox[
               TagBox[x, BoxForm`ChemistryTag[HoldComplete[mol]]], y,
                              ContentSelectable → False,
                              Selectable → False,

               DefaultBaseStyle → {FrontEnd`GraphicsHighlightColor →

                    RGBColor[0.269281, 0.535948, 0.63268]}],
                         _ :↦ $Failed}
                         ]
               ],

     With[{graphics = MoleculePlot[mol, ImageSize → moleculeImageSize[mol, 0.375]]},
                    {plot = ToBoxes[graphics, fmt]},
                    Replace[plot, {HoldPattern[GraphicsBox[x_, y___]] :↦

          GraphicsBox[TagBox[x, BoxForm`ChemistryTag[HoldComplete[mol]]], y,
                              ContentSelectable → False,
                              Selectable → False,

               DefaultBaseStyle → {FrontEnd`GraphicsHighlightColor →

                    RGBColor[0.269281, 0.535948, 0.63268]}],
                         _ :↦ $Failed}
                         ]
               ]
          ]
```

Set $RBNMoleculeFormat = False to restore built-in behavior.

Pattern Reaction Paradigms

The PatternReaction function utilizes graph rewriting to model chemical reactions. There are several options for specifying a chemical reaction using PatternReaction, but the most powerful of these, by far, is a SMARTS string. SMARTS is an acronym that stands for "SMILES (Simplified Molecular Input Line Entry System) arbitrary target specification." The power of this method comes from allowing the user to be as specific or as general as they wish to be.

For example, a condensation reaction involves the formation of a bond between two hydroxyl groups, forming an ether and giving off water as a byproduct. Utilizing SMARTS allows the user to ignore all other atoms in each molecule (or the singular molecule if it is an intramolecular condensation reaction) and only specify the atoms directly relevant to the reaction:

In[ ]:= **generalCondensationReaction =**
PatternReaction["[*:1][OH1:2].[OH1:3][*:4]>>[*:1][O:2][*:4].[OH2:3]"]

Out[ ]=

SMARTS also provides the ability to specify characteristics of the atoms that are not directly affected by the reaction. For example, we could specify that one of the hydroxyls is bonded to a nitrogen atom that is present in a six-membered ring and that the other hydroxyl is bonded to a carbon atom that is present in exactly two rings, and that these two must be in two separate molecules before the reaction takes place:

In[ ]:= **specificCondensationReaction =**
PatternReaction["([N;r6:1][OH1:2]).([OH1:3][C;R2:4])>>([N:1][O:2][C:4]).([OH2:3])"]

Out[ ]=

Although PatternReaction does not show that we have specified the ring properties for the carbon and nitrogen atoms, we can test it on a series of molecules using the ApplyReaction function. Here we have four sets of molecules, all of which fit the pattern for generalConden··. sationReaction and only the fourth one fitting the pattern for specificCondensationReaction:

In[ ]:= **(patternReactionMoleculeSets = Map[Molecule,**
{{"NO", "CO"}, {"C1CCCCN1O", "CO"}, {"NO", "C1CCCC2CCCCC21O"},
{"C1CCCCN1O", "C1CCCC2CCCCC21O"}}, {2}]) // Column

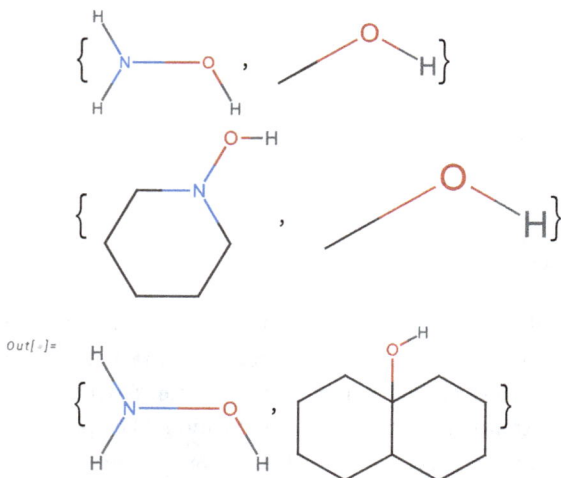

Out[ ]=

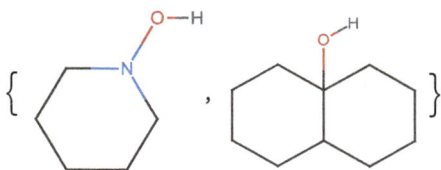

By applying generalCondensationReaction to all four sets of molecules, we expect the reactions to be carried out, and receive four sets of reactants:

```
In[ ]:= Row[{Plus@@#, "→", Plus@@ApplyReaction[generalCondensationReaction, #]},
          Spacer[9]] &/@patternReactionMoleculeSets // Column
```

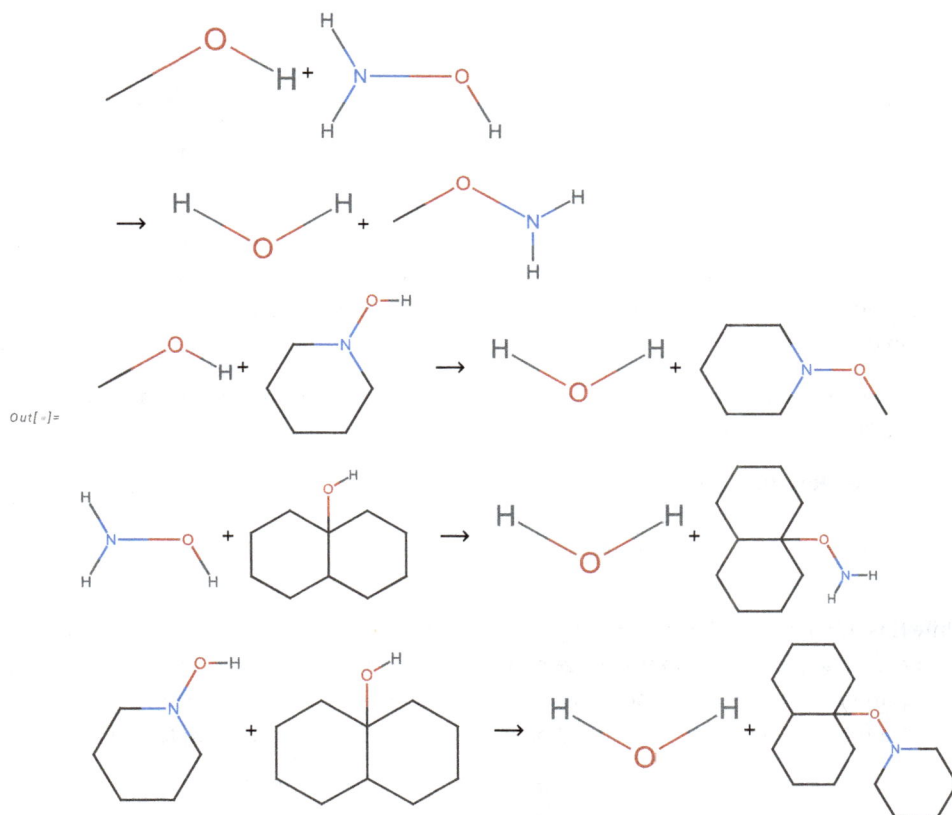

By applying specificCondensationReaction to all four sets of molecules, we expect only the fourth set to undergo the reaction, with the first three reactions each returning an empty list:

```
In[ ]:= Row[{Plus@@#, "→", Plus@@ApplyReaction[specificCondensationReaction, #] /.
          (0 → "Ø")}, Spacer[9]] &/@patternReactionMoleculeSets // Column
```

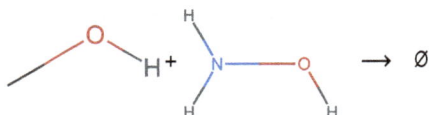

Out[ ]=

(reaction diagrams)

→ ∅

→ ∅

→

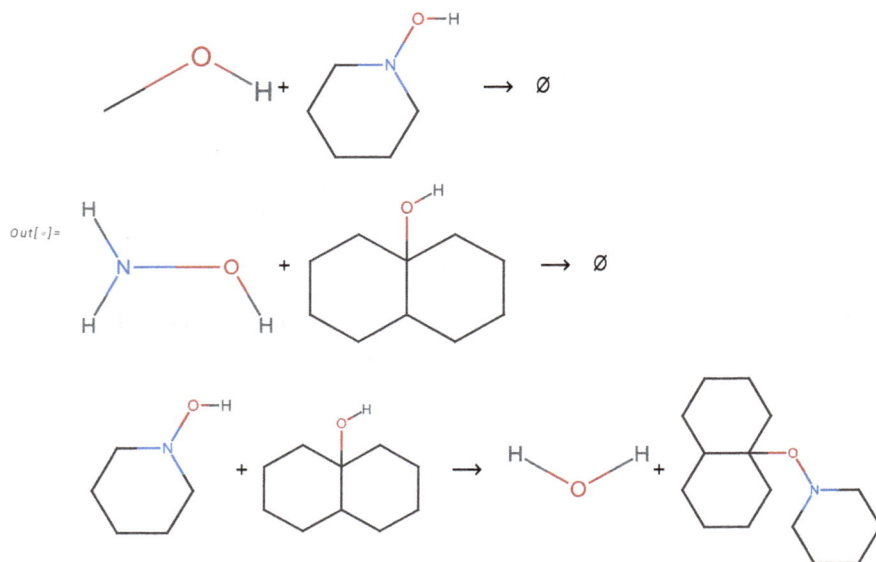

As we can see, specificCondensationReaction was applied to all four sets of reactants because each reactant contained at least one hydroxyl group. However, only the fourth set of reactants specified the requirements of specificCondensationReaction, and thus was the only reaction carried out.

Another benefit of PatternReaction is that it does not require conservation of mass or atomic identity. That is, we are able to turn lead into gold!

In[ ]:= **PatternReaction["[Pb]>>[Au]"]**

Out[ ]= Pb ⟶ Au

While this might seem silly at first, this allows for simpler specification of some reactions and can speed up applying reactions to large sets of molecules, as checking if a molecule satisfies the requirements of a SMARTS string is computationally costly. In the following sections, we explore a few paradigms of PatternReaction that utilize slightly different implementations.

Construction

The construction paradigm involves taking a starter molecule and adding onto it without specifying what is added as a reactant. This is particularly useful for building polymers using ApplyReaction, especially if we wish to restrict the binding sites for polymerization. For example, if we take a glucose molecule (note that we have not specified the stereochemistry here, although we are able to do so), we can use the construction paradigm to specify that every time an additional glucose molecule is added, its 1-carbon must attach to the 6-carbon of one of the already-present moieties:

```
In[ • ]:=  glucose = Molecule["O1C(O)C(O)C(O)C1CO"];
           glucosePolymerization = PatternReaction[
               "[C;!R:1][OH1:2]>>[C:1][O:2][C]1[C]([O])[C]([O])[C]([O])[C]([C]([O]))[O]1"];
           glucoseTetramer = ApplyReaction[glucosePolymerization, ApplyReaction[
                   glucosePolymerization, ApplyReaction[glucosePolymerization, glucose]]];
           Column@{glucose, glucosePolymerization, glucoseTetramer}
```

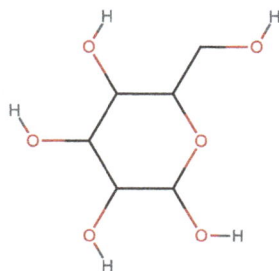

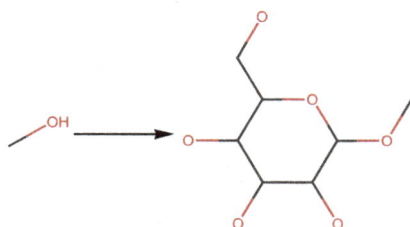

Out[•]=

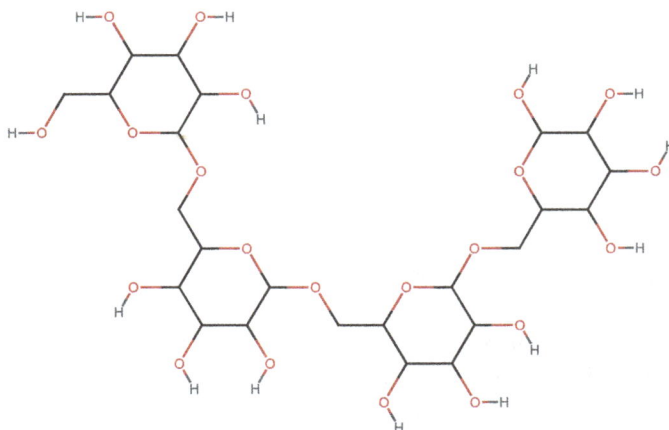

Notice that we have conveniently lost track of the resultant water molecule from this condensation reaction. This allows for less processing between polymerization steps, which would help simplify code carrying out polymerization in this way.

Self-interaction

The next PatternReaction paradigm we will discuss is self-interaction. As the name suggests, self-interaction takes two parts of the same molecule and reacts to them in some way. For example, if we take the tetramer from the previous section, there is now enough flexibility to allow for a condensation reaction to occur between two of the hydroxyls of the tetramer. A condensation reaction of this type will produce a larger ring structure. There are many ways for this reaction to occur:

```
In[ ]:=  intramolecularCondensation =
             PatternReaction["([*:1][OH:2].[OH][*:3])>>[*:1][O:2][*:3]"];
         ringClosedGlucoseTetramer =
             ApplyReaction[intramolecularCondensation, glucoseTetramer, All][[34, 1]];
         Column@{intramolecularCondensation,
             {ringClosedGlucoseTetramer, ringClosedGlucoseTetramer // MoleculePlot3D}}
```

Out[ ]=

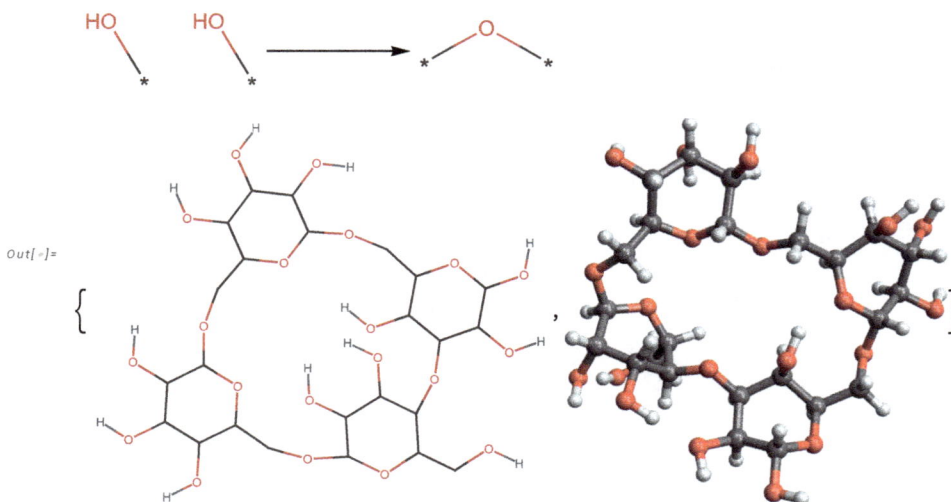

One problem that occurs with this implementation is a reaction between parts of the molecule that are too close to each other. While this is no problem for ApplyReaction, some of the resulting molecules do not make sense chemically. To remedy this problem, we can repeatedly use the wildcard character "*" in SMARTS to represent a series of unknown atoms. There are two problems with this method, however. The number of wildcards used must exactly match the number of atoms, i.e. there is no way to represent a range of unknown numbers of atoms without explicitly writing out each scenario. The other problem is that the number of wildcards corresponds to a *possible* path between two atoms, not necessarily the shortest. So between two atoms in a different molecule, there could be paths of 12, 15 and 18 atoms. Because we care about atoms being too close to each other, we may want to exclude shortest paths of length 12 or fewer, but these two atoms could still react, as there are longer possible paths between the two atoms.

Combination

The combination paradigm is the main way in which PatternReaction was originally intended to be used. In this method, we react two (or more) separate molecules to get a list of products, while maintaining atomic count for each element present. One example is that of the Diels–Alder reaction, in which a diene and a dienophile react together to form a ring, pictured here:

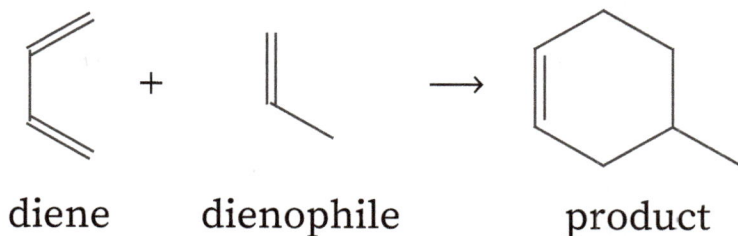

diene   dienophile   product

This reaction can be represented as follows:

In[ ]:= **PatternReaction[**
         "([C:1]=[C:2][C:3]=[C:4]).([C:5]=[C:6])>>[C:1]1[C:2]=[C:3][C:4][C:5][C:6]1"]

Out[ ]=

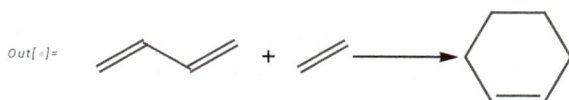

This method is especially useful in contexts of repeated reactions, which will be discussed in more detail in the following section on multiway chemical reaction networks. One of the benefits of this method is that now *every* reactant and product atom can be given a label and kept track of during the reaction. This helps reduce the possibility of losing or adding atoms unintentionally or putting atoms in the wrong place in the products. However, as alluded to earlier, specifying every atom in these reactions can be computationally costly for reactions involving much larger reactants.

Multiway Chemical Reaction Networks

Explicit reaction pathways are of great importance in chemistry for building molecules from other, more accessible or more easily manipulable molecules. Some molecules of interest require a long reaction pathway and can often be reached via multiple pathways. This feature of some reaction pathways lends itself to multiway computational analysis. Multicomputation can be used with any of the previously discussed PatternReaction paradigms, but we shall focus on the self-interaction method.

For chemical reactions, we utilize the TokenEventGraph resource function, as this allows us to visualize the reactants, products and the reactions themselves. The reactants and products are represented with tokens, and each reaction is an event in the graph (a small circle). Pictured here is a multiway graph featuring monoterpenes and several rearrangement reactions taken two layers deep:

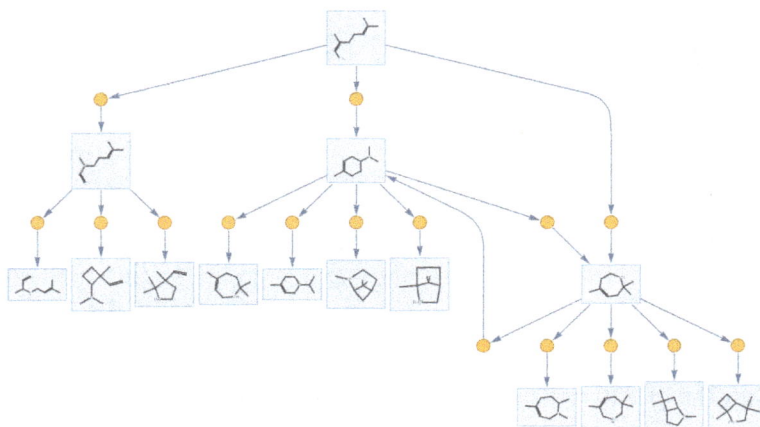

This is all wrapped in the function MultiwayChemicalReactionNetwork, which we use both to generate the set of molecules that will later be analyzed and to display the structural similarities of those molecules:

```
In[ ]:=  MultiwayChemicalReactionNetwork[reactions_,
            startingMolecules_, numSteps_, OptionsPattern[]] := Module[{},
            ResourceFunction["TokenEventGraph"][
               {{mol1_} :→ Catenate @ (Catenate[ApplyReaction[#, {mol1}, All]] & /@ reactions)},
               {startingMolecules}, numSteps,
               "EventVertices" → OptionValue["EventVertices"],
               "EventDeduplication" → OptionValue["EventDeduplication"],
               "TokenDecompositionFunction" → (DeleteDuplicates[#, MoleculeMatchQ] &),
               "TokenDeduplication" → OptionValue["TokenDeduplication"],
               "TokenEquivalenceFunction" → OptionValue["TokenEquivalenceFunction"],
               "TokenRenderingFunction" → OptionValue["TokenRenderingFunction"],
               "TokenLabeling" → OptionValue["TokenLabeling"],
               "MultiOutputEvents" → OptionValue["MultiOutputEvents"]]
            ]
```

Note that this function currently only works with reaction sets involving self-interaction reactions, so more work needs to be done to generalize this to reaction sets with varying numbers of reactants.

In order to work with the molecules generated by MultiwayChemicalReactionNetwork, we extract the molecules from the multiway graph and access the "SMILES" attribute, which returns the SMILES string for each molecule:

```
In[ ]:=  GetMultiwayChemicalReactionNetworkSMILES[multiwayGraph_] :=
            Module[{reactedMolecules, reactedMoleculeSMILES},
               #["SMILES"] & /@ VertexList @ multiwayGraph
            ]
```

This also provides a convenient and data-efficient format for storing all of the molecules generated by the multiway graph.

Chemical Space Network

A chemical space network is a graph where each node corresponds to an individual molecule and two nodes are connected by an edge if they have a similarity score (ranging between 0 and 1 inclusive) above a specified threshold. Depending upon the context, there are many possible choices for a similarity metric. All of the analysis in this post utilizes both the TopologicalFingerprint function and concepts from the Wolfram chemistry core area, and compares these fingerprints using the JaccardDissimilarity function.

The following code is the first step in creating a chemical space network. Taking the list of SMILES strings from the multiway chemical reaction network as input, this function returns an association between pairs of molecules and their similarity scores. The pairs are stored as undirected edges to make the creation of graphs later on easier:

```
In[ ]:=  ChemicalSpaceNetwork[moleculeSMILES_] := Module[{moleculeFingerprints},
             moleculeFingerprints =
                (# → TopologicalFingerprint[Molecule[#]] & /@ moleculeSMILES) // Association;
             ParallelMap[UndirectedEdge[First[#], Last[#]] →
                   N@(1 - (JaccardDissimilarity @@ {moleculeFingerprints[First[#]],
                              moleculeFingerprints[Last[#]]})) &,
                Subsets[Normal[moleculeSMILES], {2}]] // Association
          ]
```

This function is the only one in this analysis that makes use of the TopologicalFingerprint and JaccardDissimilarity functions, making it easy to implement a different similarity measure for future analyses.

Statistical Analysis

There are several statistical features of the chemical space network that are of interest to graph. The three statistics I chose to analyze were:

- The number of remaining non-singleton molecules
  - That is, the number of molecules similar to at least one other molecule at a given similarity threshold
- The number of non-singleton connected components
- The size of the largest connected component

The first and third statistics are monotonically decreasing as the similarity threshold is increased from 0 to 1. In general, we would like for the number of non-singletons to remain relatively high over the whole range of similarity thresholds, as if too many molecules are singletons, we cannot draw as many meaningful conclusions about the similarities and differences between the molecules we are analyzing.

Typically, the second statistic increases to a maximum at a similarity threshold slightly below 1 and then decreases a varying amount. However, there is a lot of variability in this statistic depending on the system of reactions and reactants chosen. For example, there is the possibility for several local maxima, and this seems to occur more often in smaller systems.

The following function takes in the association we initially created as well as a similarityIncre·. ment that controls how fine each of the plots are and returns a list of three ListLinePlot objects labeled with the appropriate information:

```
In[ ]:= ChemicalSpaceNetworkAnalysis[chemicalSpaceNetworkIn_, similarityIncrement_] :=
        Module[{chemicalSpaceNetwork = chemicalSpaceNetworkIn, graph,
            iVals = {}, numMoleculesRemaining = {}, numConnectedComponents = {},
            largestConnectedComponentSize = {}, connectedComponents},
          Do[chemicalSpaceNetwork = Select[chemicalSpaceNetwork, # ≥ i &];
            graph =
              chemicalSpaceNetwork // Keys // Graph[# &/@Union @@ List @@@ #, #] &;
            iVals = iVals ~ Join ~ {i};
            numMoleculesRemaining =
              numMoleculesRemaining ~ Join ~ {graph // VertexList // Length};
            connectedComponents = ConnectedComponents @ graph;
            numConnectedComponents =
              numConnectedComponents ~ Join ~ {connectedComponents // Length};
            largestConnectedComponentSize = largestConnectedComponentSize ~
              Join ~ {Max[{0, Length /@ connectedComponents} // Flatten]};,
            {i, 0, 1, similarityIncrement}
          ];
          {ListLinePlot[MapThread[List, {iVals, numMoleculesRemaining}],
              PlotLabel → "Non–Singletons Remaining", PlotRange → Full],
            ListLinePlot[MapThread[List, {iVals, numConnectedComponents}],
              PlotLabel → "Number of Connected Components", PlotRange → Full],
            ListLinePlot[MapThread[List, {iVals, largestConnectedComponentSize}],
              PlotLabel → "Size of Largest Connected Component", PlotRange → Full]}
        ]
```

Here are the plots generated for the same monoterpene system mentioned in the "Multiway Chemical Reaction Networks" section, this time carried out four layers deep:

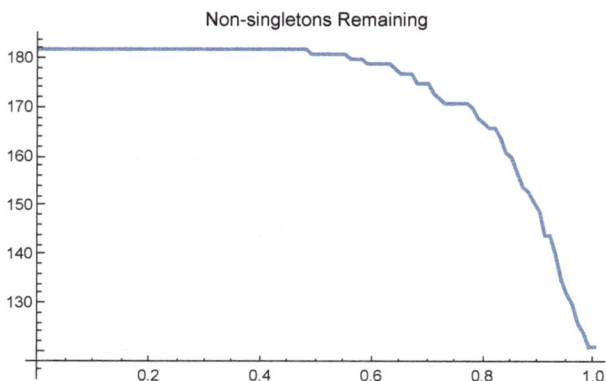

We can see that at a similarity threshold of 1, i.e. molecules are only similar if they are topologically identical (according to their topological fingerprints), approximately one-third of the molecules are singletons. The majority of the chemical reaction systems I investigated had between 10% and 30% of their molecules as singletons at a similarity threshold of 1:

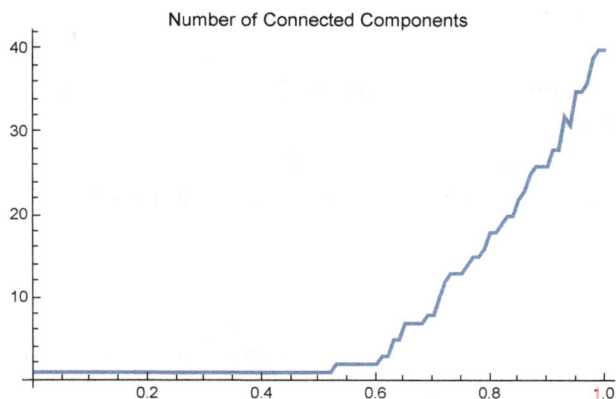

Number of Connected Components

Here we can see that the number of connected components does not drop off as the similarity threshold approaches 1. We can also see that the chemical space network remains connected until we impose a similarity threshold slightly over 0.5 and then steadily grows to around 40 clusters:

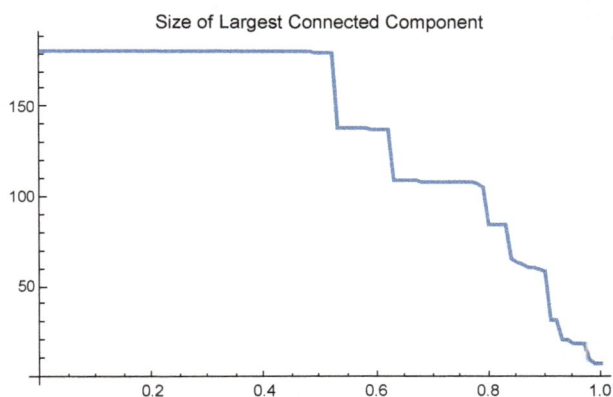

Size of Largest Connected Component

This graph is very similar to a step function, as we can see several thresholds where the largest connected component fractionates into two or more clusters. Another behavior that can be seen with the size of the largest connected component is a steep dropoff at a high similarity threshold. Based on my investigation, this seems to be more common in polymer-building systems.

Graphs

We would also like to display the chemical space network at various similarity thresholds to aid in our analysis of these systems. The following function creates a list of chemical space network graphs, also including the ability to click any of the nodes to get the SMILES string of its associated molecule:

```
In[ ]:=  ChemicalSpaceNetworkGraphs[moleculeSMILES_, chemicalSpaceNetworkEdges_,
            {minSimilarityThreshold_, maxSimilarityThreshold_, thresholdIncrement_}] :=
          Module[{graphs = {}, edges = chemicalSpaceNetworkEdges},
            Do[
              edges = Select[edges, # ≥ i &];
              graphs =
                graphs ~ Join ~ {Graph[Button[#, CopyToClipboard["\"" <> # <> "\""]] & /@
                  Normal[moleculeSMILES], Keys @ edges,
                  GraphLayout → "GravityEmbedding", PlotLabel → i]};,
              {i, minSimilarityThreshold, maxSimilarityThreshold, thresholdIncrement}
            ];
            graphs
          ]
```

Pictured here are the chemical space networks for the four-layer-deep monoterpene system described earlier at six different similarity thresholds:

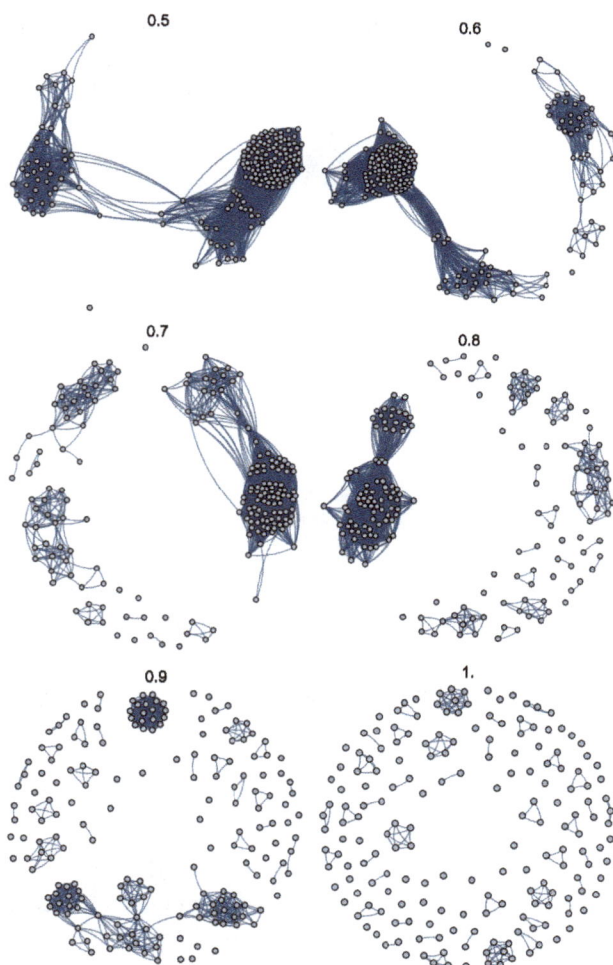

Combining the Two

Colored Clusters on the Multiway Graph

One way we can combine the multiway chemical reaction network and chemical space network is to display the clusters (connected components) from the chemical space network on the multiway graph using a unique color for each cluster.

Although the only difference between this graph and the original multiway graph is the coloring, we create it separately, as it is much easier to modify this way. The ClusterColored MultiwayChemicalReactionNetwork function also allows us to only color clusters of a given size or larger. Any clusters too small to be colored are given a gray background.

Shown here is the three-layer-deep monoterpene system analyzed at a similarity threshold of 0.9:

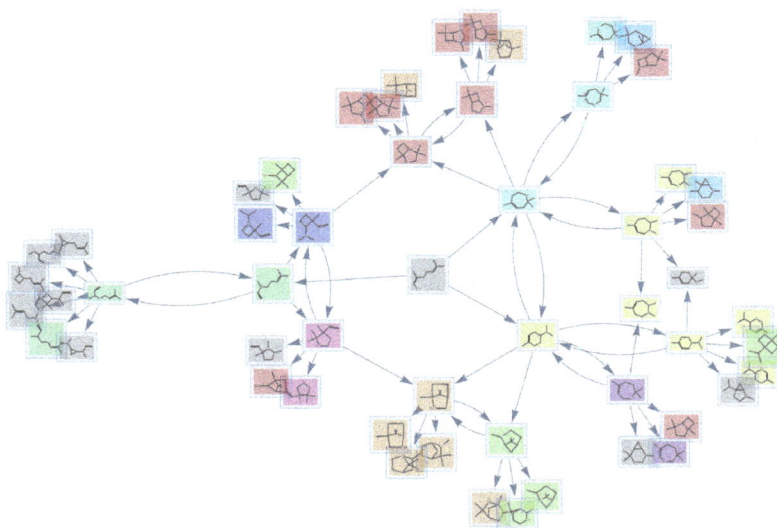

We can see that, for the most part, molecules of the same cluster (those with the same colors) are genealogically close. There are some notable exceptions, such as light green, which has only five molecules in its cluster, but those molecules are scattered across the graph and are not very genealogically close to each other.

Genealogical Distance on Chemical Space Networks

One other way we can combine the information from the multiway graph and the chemical space network is using the distance between the nodes in the multiway graph, which will be referred to as the genealogical distance. For each cluster in the chemical, one node is chosen as the central node, and the genealogical distances from that central node to all other nodes in the cluster are calculated. Each node is then given a color corresponding to its genealogical distance from its central node, with central nodes all colored red. The redder a node is, the closer it is genealogically to its cluster's central node.

Shown here is the four-layer-deep monoterpene system at a similarity threshold of 0.9 with the genealogical distance indicated by the colors:

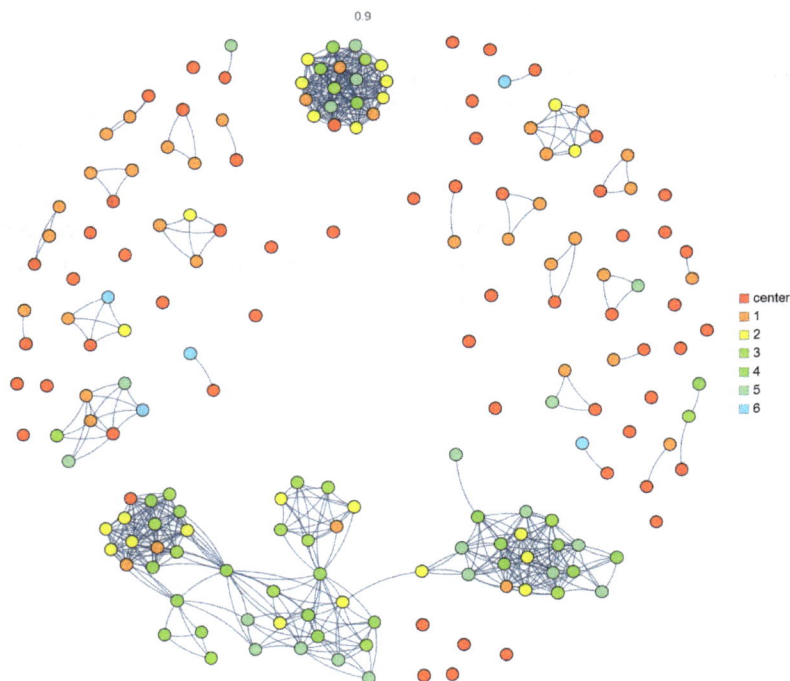

As we can see in the image, most of the nodes within smaller clusters are genealogically very close to each other, with only a few of the smaller clusters having colors besides red and orange. For the larger cluster, we see more yellow and green, which shows that the molecules within the clusters are more genealogically spread out. It is possible that these clusters would appear redder if we allowed for genealogical distance to go down through the multiway graph, i.e. calculate the distance through "descendant" molecules, but for most analytical purposes, the definition of genealogical distance used in this analysis seems more applicable.

Branchial Analysis

Because we are investigating the genealogical distance between molecules in these reaction networks, branchial graphs are of much interest. The BranchialGraphs resource function provides a list of graphs where two nodes are connected if they share a "parent" node, i.e. a node such that there exists a directed edge from the "parent" node to the two nodes of interest (called "child" nodes).

Shown here are the branchial graphs for a two-layer-deep Diels–Alder reaction network discussed in the "Pattern Reaction Paradigms" section:

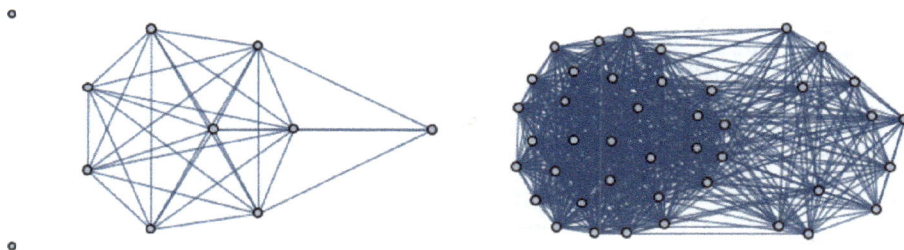

Even going only two layers deep, we can see the beginnings of a cluster separation in the genealogy of the molecules. Each side of the branchial graph on the right has many more internal connections than connections with the other side.

Unfortunately, the BranchialGraphs function only works on acyclic graphs, so any multiway chemical reaction network that has reversible or cyclic reactions cannot be analyzed using branchial graphs.

Full Analysis

Example: Geranyl Cation/Monoterpene Biosynthesis Reactions (Three Steps)

Statistical analysis graphs:

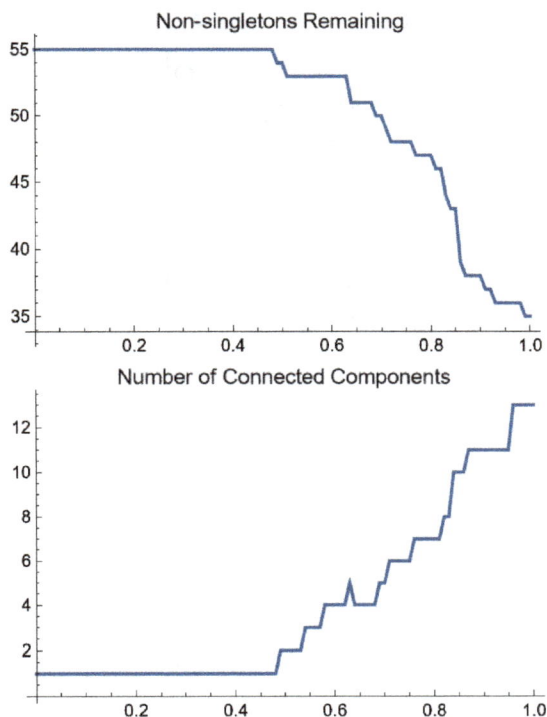

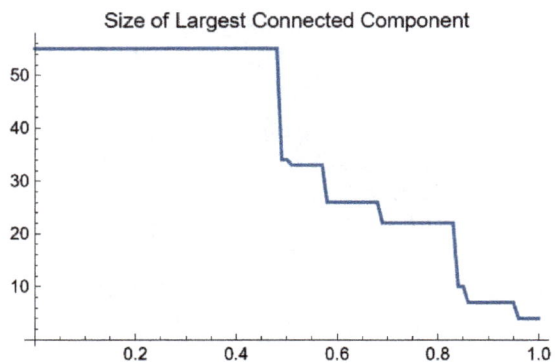

Size of Largest Connected Component

Chemical space network at a similarity threshold of 0.9:

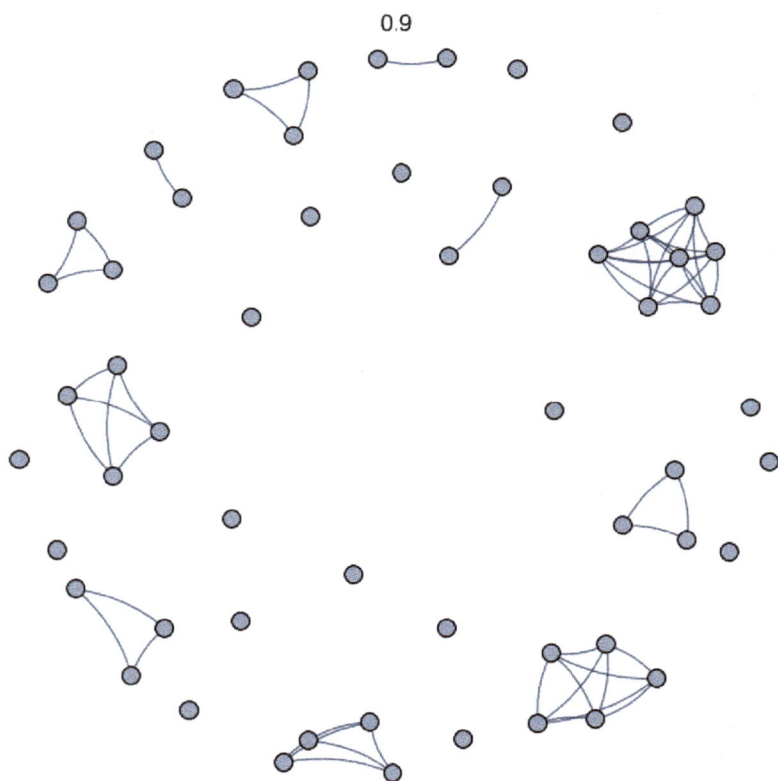

0.9

Multiway chemical reaction network colored using the clusters shown previously:

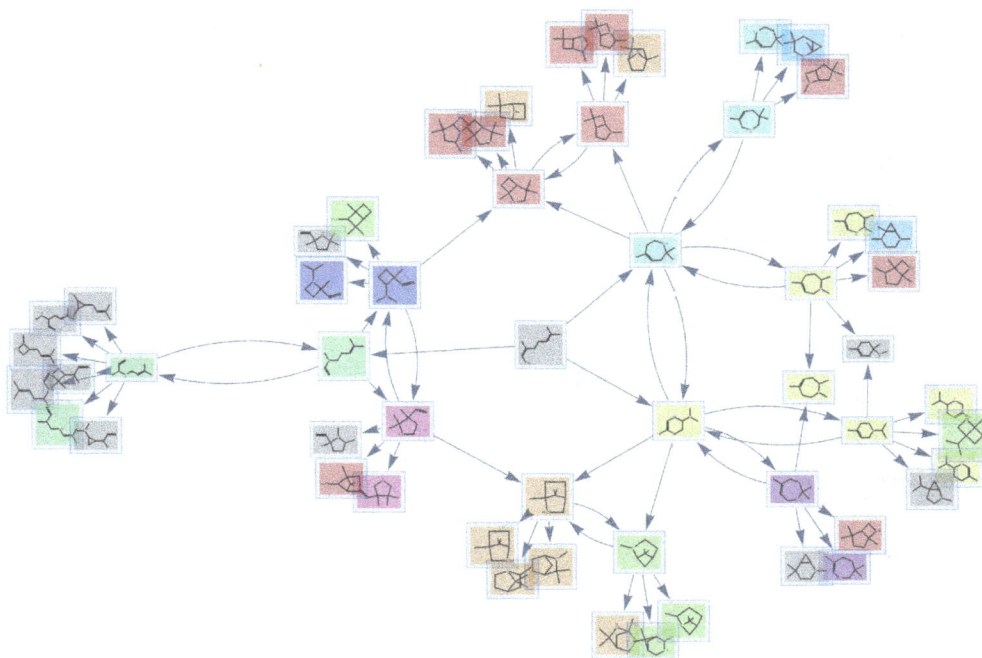

Chemical space network with nodes colored according to genealogical distance from cluster centers:

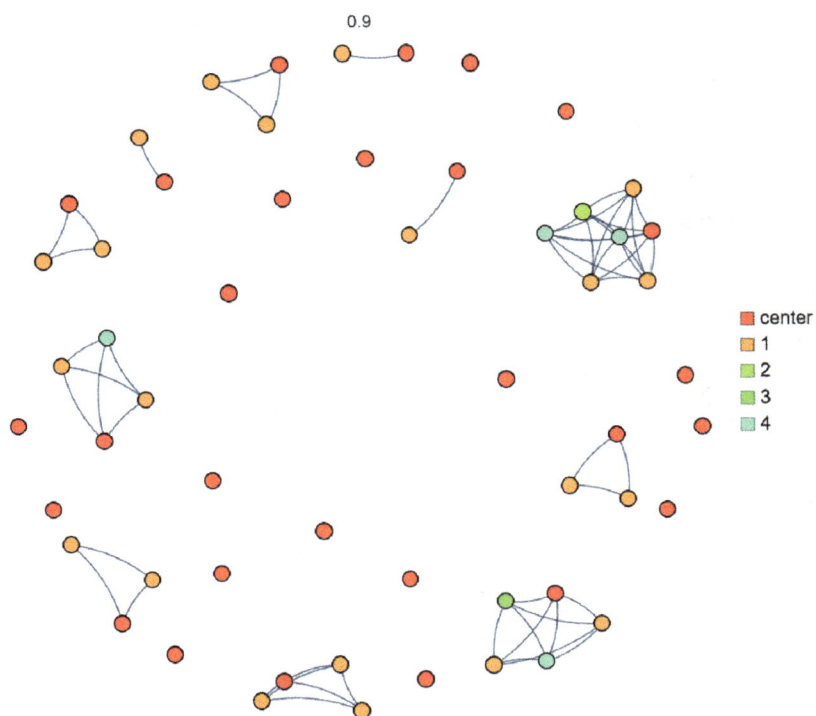

Principal component analysis on genealogical distance matrix from multiway graph:

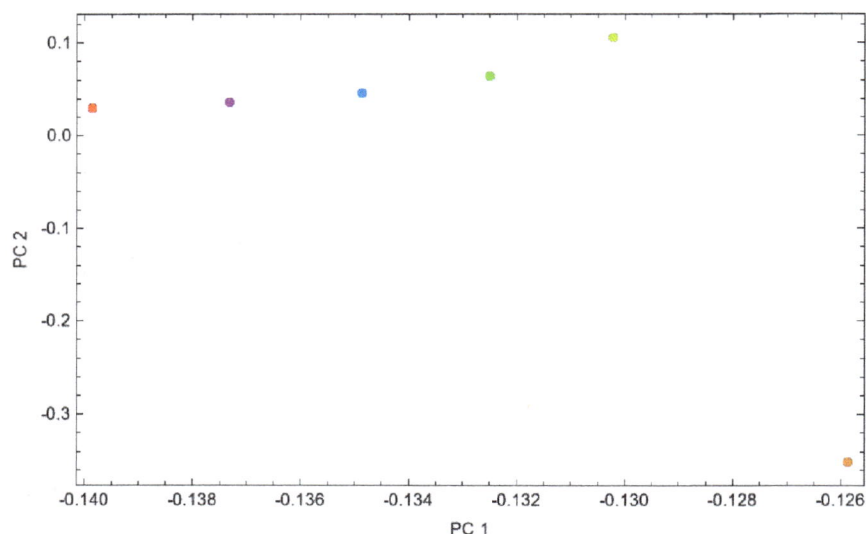

Code

This section is a potential workflow to generate the analysis for a multiway chemical reaction system of your choice.

The variables in this input cell are for:

- fullAnalysisMolecules: list of molecules to start your multiway chemical reaction system

- fullAnalysisReactions: list of reactions for your system

- fullAnalysisNumSteps: the number of reaction steps your system will undergo

- fullAnalysisSMILES: the number of reaction steps your system will undergo

```
In[ ]:=  fullAnalysisMolecules = {geranylCation};
         fullAnalysisReactions = monoTerpeneBiosynthesisReactions;
         fullAnalysisNumSteps = 4;
         fullAnalysisSMILES =
             GetMultiwayChemicalReactionNetworkSMILES[MultiwayChemicalReactionNetwork[
                 fullAnalysisReactions, fullAnalysisMolecules, fullAnalysisNumSteps]];
```

The variable in this input cell is for:

- fullAnalysisChemicalSpaceNetwork: association of molecule pairs to their similarity scores

```
In[ ]:=  fullAnalysisChemicalSpaceNetwork = ChemicalSpaceNetwork[fullAnalysisSMILES];
         ChemicalSpaceNetworkAnalysis[fullAnalysisChemicalSpaceNetwork, 0.01]
```

The variable in this input cell is for:

- fullAnalysisChemicalSpaceNetworkGraphs: chemical space network graphs at specified similarity thresholds

In[ ]:= **fullAnalysisChemicalSpaceNetworkGraphs = ChemicalSpaceNetworkGraphs[**
      fullAnalysisSMILES, fullAnalysisChemicalSpaceNetwork, {0.5, 1, 0.05}]

The variables in this input cell are for:

- fullAnalysisMinimumClusterSize: smallest cluster size that will be given a color on the multiway graph

- fullAnalysisChemicalSpaceNetworkGraph: chemical space network chosen from fullAnalysisChemicalSpaceNetworkGraphs to take clusters from

- fullAnalysisMultiwayGraph: multiway graph colored according to clusters of fullAnalysisChemicalSpaceNetworkGraph

In[ ]:= **fullAnalysisMinimumClusterSize = 3;**
    fullAnalysisChemicalSpaceNetworkGraph =
      fullAnalysisChemicalSpaceNetworkGraphs〚9〛;
    fullAnalysisMultiwayGraph = ClusterColoredMultiwayChemicalReactionNetwork[
      fullAnalysisReactions, fullAnalysisMolecules,
      fullAnalysisNumSteps, fullAnalysisChemicalSpaceNetworkGraph,
      fullAnalysisMinimumClusterSize, "EventVertices" → False];

This input cell returns a chemical space network with the nodes colored according to their genealogical distance from the central node of their respective connected components. Note the 0.9 *must* match the similarity threshold of fullAnalysisChemicalSpaceNetworkGraph:

In[ ]:= **ChemicalSpaceNetworkAndGenealogicalDistanceGraph[fullAnalysisSMILES,**
    fullAnalysisChemicalSpaceNetwork, fullAnalysisMultiwayGraph, 0.9]

This input cell returns a scatter plot of the principal components analysis on the genealogical distance matrix generated from the multiway chemical reaction network:

In[ ]:= **ColoredDimensionReducedClusters[fullAnalysisSMILES,**
    fullAnalysisChemicalSpaceNetworkGraph, fullAnalysisMinimumClusterSize, 2]

Future Work

Branchial Analysis

As mentioned previously, branchial analysis as it currently exists in Wolfram Language is incompatible with most reaction networks due to the existence of directed cycles. It would be an interesting endeavor to create a version of BranchialGraphs that is able to work with cyclic graphs. I suspect this would have applications far beyond the analysis done for this project, so I look forward to seeing someone tackle this problem.

Multiway Chemical Reaction Network

Number of Reactants

As of now, MultiwayChemicalReactionNetwork can only handle reaction sets that have the same number of reactants for all reactions, and even then the function has to be modified manually to accommodate different numbers of reactants. A logical next step would be to rewrite the rules given to the token event graph function in such a way as to allow for a set of reactions with different numbers of reactants.

Pruning Function

Many reaction systems of interest quickly explode with the number of possible molecules, slowing any analysis beyond the point of usefulness. To remedy this, a function that removes molecules at each step in the token event graph based on some input criteria would allow for larger-depth investigations of some of these systems.

There could also be a Random option for the pruning function that only allows the multiway graph to keep a certain number of molecules at each time step, choosing these at random.

Probabilistic Multiway Computation

One of the downsides of using multiway computation to model chemical reaction networks is that the edges in the generated graphs all have the same weights, which can be interpreted as the reactions being equiprobable. This is not very realistic, and it would be of great use to incorporate a probabilistic aspect into the multiway computation. For example, if each reaction was assigned a probability of occurring, each molecule could be paired with a list of probabilities associated with how likely it is to reach that molecule from a given state after a given number of steps. The greatest challenge with this would likely be visualization, so a great deal of creativity would be needed to incorporate this into Wolfram Language.

Chemical Space Network

Other Similarity Metrics

The analysis with the current implementation can only be done using the topological fingerprints and the JaccardDissimilarity function for similarity comparisons. To expand the analytical abilities of this method, a more general form for the ChemicalSpaceNetwork function allowing for different metrics to be used should be implemented

GUI

A Wolfram Summer School 2023 project by Joshua Khorsandi explored chemical space networks in the context of comparing bio activity to chemical structure. As part of his project, he developed a GUI that allowed for individual cluster investigation and molecular inspection, among other capabilities [4]. His work should be modeled after and expanded upon to incorporate the multiway chemical reaction network and analyses described in this post.

Concluding Remarks

The versatility of multiway computation has great potential in chemical analysis, especially when paired with more traditional analysis methods, such as chemical space networks. From the results presented, it is clear that multiway chemical reaction networks and chemical space networks provide very similar yet distinct information about the chemical systems that they are analyzing. This makes pairing them together a very powerful tool, verifying the validity of each other while also providing more information than either method could on its own.

Acknowledgments

I would like to acknowledge Robert Nachbar, who served as my mentor throughout this project. He consistently helped me to focus my project toward goals that were both significant and achievable in the timeframe of the Wolfram Summer School and to swiftly resolve any coding challenges I ran into during my project. I would also like to thank Stephen Wolfram and Mads Bahrami for directing me toward this project and providing their insights.

References

1. G. M. Maggiora and J. Bajorath (2014), "Chemical Space Networks: A Powerful New Paradigm for the Description of Chemical Space," *Journal of Computer-Aided Molecular Design*. pubmed.ncbi.nlm.nih.gov/24925682.

2. A.-D. Vlad (2023), "[FSWS23] A Multiway Model of Chemical Reactions," *Wolfram Community*. community.wolfram.com/groups/-/m/t/2778062.

3. R. Nachbar (2022), "Modeling Chemical Reaction Networks with Token-Event Graphs," talk given at the Wolfram Technology Conference 2022. www.wolfram.com/broadcast/video.php?c=104&p=6&disp=list&v=3812.

4. J. Khorsandi (2023), "Chemical Space Networks from Molecule Data," *Wolfram Community*. community.wolfram.com/groups/-/m/t/2958091.

Access the Full Code

Scan or visit wolfr.am/WSS2024-Frieler.

Cite This Notebook

"Analyzing Multiway Chemical Reaction Networks Using Chemical Space Networks"
by Nicholas Frieler
Wolfram Community, STAFF PICKS, July 9, 2024
community.wolfram.com/groups/-/m/t/3210073

Electronic Structure of Atoms Using Quantum Computing Techniques

MICHAŁ ZDZIENNICKI

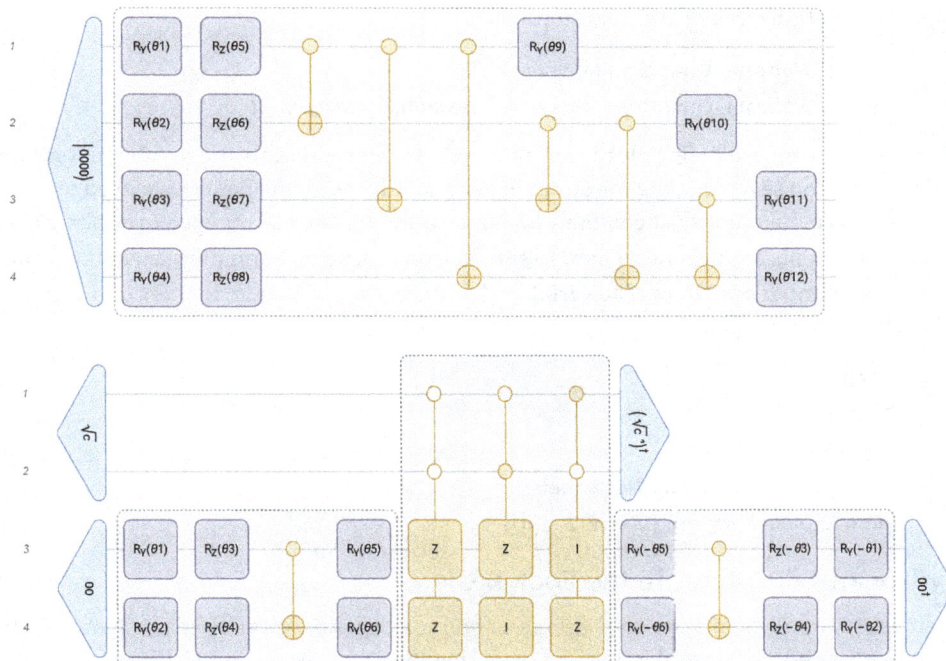

This project aims to simulate the ground state of hydrogen-like atoms using quantum computing tools in Mathematica. It employs quantum circuits to represent fermionic operators, following the principles of the Hartree–Fock theory. By applying mixed quantum-classical algorithms like the variational quantum eigensolver, the goal is to minimize energy within the circuits, thereby transforming initial states into the ground state of a Hartree–Fock Hamiltonian. This research demonstrates the potential of quantum computing in simulating atomic structures and computational chemistry.

Helium Atom as a Quantum Circuit

Creating a Computational Basis for the Problem

Required Imports and Global Variables

First, we import required packages. We will use the Wolfram Quantum Framework to simulate quantum circuits and quantum states:

```
In[ ]:=  Needs["Wolfram`QuantumFramework`"];
         Needs["Wolfram`QuantumFramework`ExampleRepository`"]
```

In the second step, we define global variables. NChannels represents the number of orbitals that we want to use to simulate the atom; in general, the more, the better. We need to, however, keep the number of channels low since they will later be mapped to qubits. Large numbers of qubits are in general hard to simulate on a classical computer. $Z = 2$ stands for the atomic number of helium. The variable e0 is the energy of helium in units of the ground-state energy of hydrogen:

```
In[ ]:=  NChannels = 4;
         Z = 2;
         e0 = -0.5;
         init = ConstantArray[0, 3 NChannels];
         init = ReplacePart[init, {1 → π, 2 → π}]
```

Hydrogen-Like States in the Fock Basis

To simulate simple atoms, we can use one-to-one correspondence of electron algebra with qubit algebra. We define each qubit as a representation of a hydrogen-like orbital. Let us use four orbitals (two 1s and two 2s) for the sake of this helium atom calculation:

```
In[ ]:=  tableOfStates = Flatten[#, 3] &[ Table[{n → nIter, l → lIter, m → mIter, ms → msIter},
              {nIter, 1, Quotient[NChannels, 2]+1}, {lIter, 0, nIter-1},
              {mIter, -lIter, lIter}, {msIter, -1/2, 1/2, 1}]];
         tableOfStates = Take[tableOfStates, NChannels];
         TableForm[#, TableHeadings → {Range[tableOfStates // Length], None}] &[
             tableOfStates]
```

Out[ ]//TableForm=

1	$n \to 1$	$l \to 0$	$m \to 0$	$ms \to -\frac{1}{2}$
2	$n \to 1$	$l \to 0$	$m \to 0$	$ms \to \frac{1}{2}$
3	$n \to 2$	$l \to 0$	$m \to 0$	$ms \to -\frac{1}{2}$
4	$n \to 2$	$l \to 0$	$m \to 0$	$ms \to \frac{1}{2}$

Hartree Integrals

The general Hartree–Fock Hamiltonian can be described in Fock basis in the form

$$H = \sum_{i,j=0}^{N} h_{ij}\, a_i^\dagger a_j + \frac{1}{2} \sum_{i,j,k,l=1}^{N} h_{ijkl}\, a_i^\dagger a_j^\dagger a_k a_l,$$

where

$$h_{ij} = \int d\tau\, \phi_i^*(r)\left(\frac{p^2}{2m} + V(r)\right)\phi_j(r)$$

$$h_{ijkl} = \int\int d\tau_1\, d\tau_2\, \phi_i^*(r)\,\phi_j^*(r')\, V(r-r')\,\phi_k(r')\,\phi_l(r)$$

are Hartree integrals. In the case of the helium atom, h_{ij} is the diagonal for hydrogen-like orbitals and its value is $h_{nn} = \frac{Z^2 E_0}{n^2}$, where n is the principal quantum number. The two-body potential is given by the formula $V(r-r'') = \frac{1}{|r-r'|}$. Our choice of basis orbital functions makes the problem clear in terms of physical meaning of the transition terms. It is, however, not the best in terms of computation. Orbital two-body integrals in the hydrogen basis give accurate values for transition probability. Although accurate, they are numerically heavy. One may take other bases into consideration, such as STO-3G. Hydrogen-like integrals can be computed semi-analytically. The integration is cumbersome and can be found in the appendix.

This bit of code takes up to two minutes to compute:

```
In[*]:= Y[{li_?IntegerQ, mi_?IntegerQ}, {lh_?IntegerQ, mh_?IntegerQ},
        {lj_?IntegerQ, mj_?IntegerQ}, {lk_?IntegerQ, mk_?IntegerQ},
        {l_?IntegerQ, m_?IntegerQ}] := (-1)^m Sqrt[(2 lk+1)/(2 lh+1) (2 lj+1)/(2 li+1)] ×
        ClebschGordan[{l, 0}, {lk, 0}, {lh, 0}] × ClebschGordan[{l, m}, {lk, mk}, {lh, mh}] ×
        ClebschGordan[{l, 0}, {lj, 0}, {li, 0}] × ClebschGordan[{l, -m}, {lj, mj}, {li, mi}];
```

```
In[*]:= Rad[n_?IntegerQ, l_?IntegerQ, r_] :=
        Sqrt[(Z Factorial[(n-l-1)])/(n^2 Factorial[n+l])] × Exp[-Z r/n] ((2 Z)/n r)^(l+1) LaguerreL[n-l-1, 2 l+1, (2 Z r/n)] 1/r;
```

```
In[*]:= R[{ni_?IntegerQ, li_?IntegerQ}, {nh_?IntegerQ, lh_?IntegerQ},
        {nj_?IntegerQ, lj_?IntegerQ}, {nk_?IntegerQ, lk_?IntegerQ}, l_?IntegerQ] :=
        NIntegrate[x^2 Rad[ni, li, x] × Rad[nj, lj, x] *
        (1/x^(l+1) Integrate[Rad[nh, lh, y] × Rad[nk, lk, y] y^(l+2), {y, 0, x}] +
        x^l Integrate[Rad[nh, lh, y] × Rad[nk, lk, y]/y^(l-1), {y, x, ∞}]), {x, 0, ∞}];
```

```
In[*]:= Hartree[{ni_?IntegerQ, li_?IntegerQ, mi_?IntegerQ},
        {nh_?IntegerQ, lh_?IntegerQ, mh_?IntegerQ},
        {nj_?IntegerQ, lj_?IntegerQ, mj_?IntegerQ},
        {nk_?IntegerQ, lk_?IntegerQ, mk_?IntegerQ}] := Module[{i, f},
        i = Max[Abs[lk-lh], Abs[li-lj]];
        f = Min[lk+lh, li+lj];
```

```
     Sum[
       If[Y[{li, mi}, {lh, mh}, {lj, mj}, {lk, mk}, {l, m}] == 0, 0, Y[{li, mi}, {lh, mh},
              {lj, mj}, {lk, mk}, {l, m}]*N[R[{ni, li}, {nh, lh}, {nj, lj}, {nk, lk}, l]]],
       {l, i, f}, {m, -l, l}]];
```

In[ ]:= sStates =
```
     Table[{i, j, k, l} → Hartree[{i, 0, 0}, {j, 0, 0}, {k, 0, 0}, {l, 0, 0}], {i, 1, 2}, {j, 1, 2}, {k, 1, 2},
              {l, 1, 2}] // ArrayFlatten // Flatten // Association
```

Out[ ]= <|{1, 1, 1, 1} → 1.25, {1, 1, 1, 2} → 0.17871, {1, 2, 1, 1} → 0.17871, {1, 2, 1, 2} → 0.419753,
       {1, 1, 2, 1} → 0.17871, {1, 1, 2, 2} → 0.0438957, {1, 2, 2, 1} → 0.0438957,
       {1, 2, 2, 2} → 0.0171633, {2, 1, 1, 1} → 0.17871, {2, 1, 1, 2} → 0.0438957,
       {2, 2, 1, 1} → 0.0438957, {2, 2, 1, 2} → 0.0171633, {2, 1, 2, 1} → 0.419753,
       {2, 1, 2, 2} → 0.0171633, {2, 2, 2, 1} → 0.0171633, {2, 2, 2, 2} → 0.300781|>

Defining the Problem as a Quantum Circuit

Computational Ansatz

To conduce any variational method computation, we need a state ansatz. In the case of quantum computing, the ansatz is a quantum circuit that is controllable and transforms a series of qubit 0s to an arbitrary vector in Hilbert space. For this calculation, we chose to use the truncated two-local ansatz, which consists of four gate layers. The first and second layers are parametrized rotations on the Bloch sphere over the y and z axes subsequently. These layers generate an arbitrary non-entangled state. Next we have a sequence of CNOT gates that aims to entangle the state created in the first two layers. Finally, we transfer this state to a different basis. Originally, this is achieved by using two rotations over the x and y axes. We choose to truncate the last layer. The ansatz can be thought of as a state that is first prepared in basis A, then entangled and measured in basis B:

In[ ]:= Options[parametrizedLayer] = {"Symbol" → "θ"};
```
     parametrizedLayer[op_, range_List, OptionsPattern[]] :=
        Sequence@@Table[op[Symbol[OptionValue["Symbol"] <> ToString[i]]] →
              Flatten[Position[range, i]], {i, range}]

     parametrizedLayer["RY", Range[NChannels+1, 2 NChannels]];
```

In[ ]:= entaglementLayer[cop_, range_List] :=
```
        Sequence@@Thread[cop → Subsets[Range[NChannels], {2}]]

     entaglementLayer["CNOT", Range[NChannels]];
```

In[ ]:= NLayers = 3;

In[ ]:= params = Table[Symbol["θ" <> ToString[i]], {i, 1, NLayers*NChannels}]

Out[ ]= {θ1, θ2, θ3, θ4, θ5, θ6, θ7, θ8, θ9, θ10, θ11, θ12}

```
In[•]:= ansatz = QuantumCircuitOperator[{
            parametrizedLayer["RY", Range[NChannels]],
            parametrizedLayer["RZ", Range[NChannels + 1, 2 NChannels]],
            entaglementLayer["CNOT", Range[NChannels]],
            parametrizedLayer["RY", Range[1 + 2 NChannels, 3 * NChannels]]
            }, "Parameters" → params, "Label" → "2–Local Ansatz"];
```

```
In[•]:= ansatz["Diagram"]
```

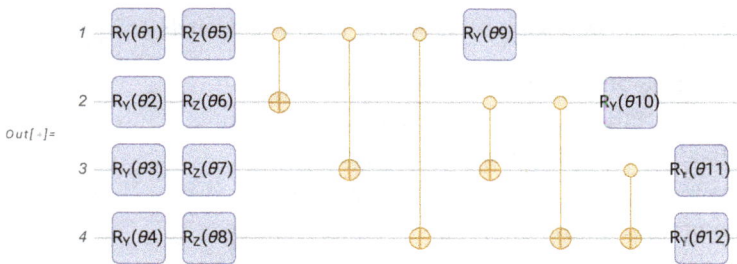

```
In[•]:= stateAnsatz = ansatz[];
```

```
In[•]:= QuantumCircuitOperator[{StringRepeat["0", NChannels], ansatz}]["Diagram"]
```

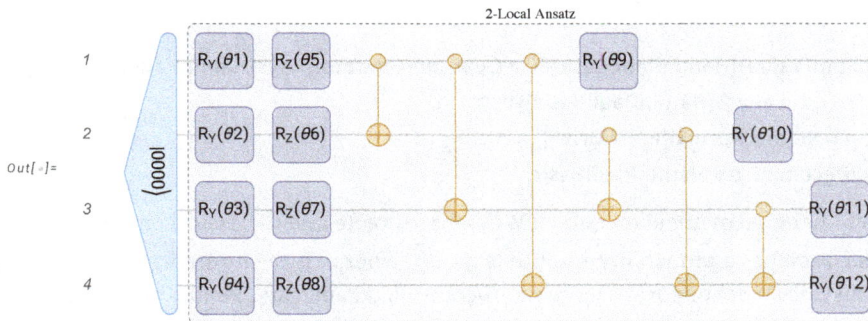

```
In[•]:= ket = stateAnsatz["StateVector"];
        bra = stateAnsatz["Dagger"]["StateVector"];
```

Jordan–Wigner Transformation of a Hamiltonian

To obtain the computational Hamiltonian, we need to transform the fermionic creation and annihilation operators acting on atomic orbitals to the sequence of gates acting on the computational qubits. This is called the Jordan–Wigner transformation [5]:

```
In[•]:= z = Table[QuantumOperator["Z", {i}], {i, 1, NChannels}];
        aFerm = Table[QuantumTensorProduct[Join[Take[z, n − 1],
                    {QuantumOperator[1/2 ("X" + i "Y"), {n}]}]], {n, 1, NChannels}];
        aFermDag = Table[aFerm[[n]]["Dagger"], {n, 1, NChannels}];
        nFerm = Table[aFermDag[[i]] @ aFerm[[i]], {i, 1, NChannels}];
```

The Hamiltonian as a Quantum Circuit

The Hamiltonian as a non-unitary operator doesn't have trivial quantum circuit representation. One needs to use some method to simulate it. For the sake of this computation, we use a method of decomposing the Hamiltonian into the sum of tensor products of Pauli gates. Then we compile a multiplexer to create a sum weighted by the ancilla qubits. We start with the definition of the Hartree Hamiltonian in the computational basis (the computation may take up to two minutes):

```
In[ ]:=  Ham = Sum[((Z^2 e0 1/n^2) /. tableOfStates[[i]]) * nFerm[[i]], {i, 1, NChannels}] + 2 * Sum[
             sStates[{n /. tableOfStates[[i]], n /. tableOfStates[[j]],
                 n /. tableOfStates[[k]], n /. tableOfStates[[l]]}] ×
                 aFermDag[[i]] @ aFermDag[[j]] @ aFerm[[k]] @ aFerm[[l]],
             {j, 1, NChannels}, {i, 1, j}, {k, 1, NChannels}, {l, 1, k}]
         HamComp = QuantumOperator[Ham, "IIIII"];
         HamDecomposed = HamComp["PauliDecompose"]
```

We decompose the Hamiltonian in the Pauli basis and truncate small terms, as they result from computational errors:

```
In[ ]:=  HamInPauliBasis = Select[HamDecomposed, Abs[#] > .01 &];
         H =
             Sum[Values[HamInPauliBasis][[i]] × QuantumOperator[Keys[HamInPauliBasis][[i]]],
                 {i, 1, Length[HamInPauliBasis]}];
         mean = Re[bra.Normal[H["Matrix"]].ket];
         pauliDecompose = HamInPauliBasis;
```

Now we define the set of ancillary qubits. We want to create a semiclassical circuit in which each ancilla qubit has a probability amplitude of $\sqrt{c}$, where c is the numerical coefficient of a given Pauli gate. We then form a large multi-controlled Pauli gate. Finally, we join all of the Pauli gates:

```
In[ ]:=  ancillary = QuantumState[Sqrt @ Values[pauliDecompose], "Label" → " √c "];
         L = ancillary["Qudits"];
```

```
In[ ]:=  circuitH = QuantumCircuitOperator[
             {
                 multiplexer
             }, "H−Circuit"
         ];
         circuitH["Diagram", ImageSize → Full]
```

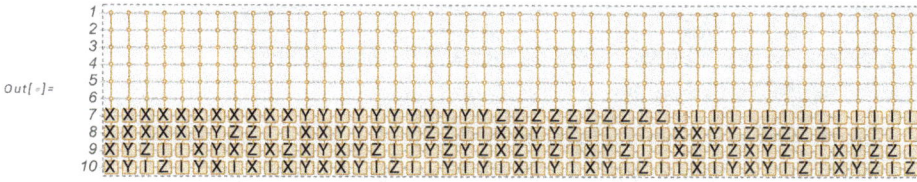

To compile the whole circuit, we need to add the ansatz circuit in the beginning and the inverse ansatz at the end. Such a circuit computes the expected value of energy:

```
In[ ]:= AppliedHCircuit = QuantumCircuitOperator[{
            QuantumState[StringRepeat["0", NChannels]] → Range[NChannels]+L,
            ansatz → Range[NChannels]+L,
            ancillary,
            circuitH,
            ansatz["Dagger"] → Range[NChannels]+L,
            QuantumState[StringRepeat["0", NChannels]]["Dagger"] →
              Range[NChannels]+L,
            ancillary["Conjugate"]†
          }];
        AppliedHCircuit["Diagram", ImageSize → Full, Expand → 2]
```

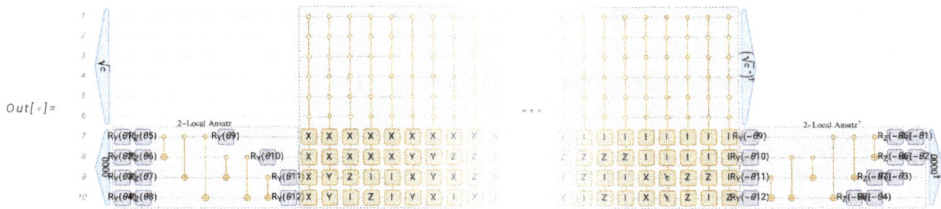

We can make the simple circuit if we use the ansatz state in an explicit form:

```
In[ ]:= ExpectedValueCircuit = QuantumCircuitOperator[{
            QuantumState[stateAnsatz, "Label" → "ψ(θ)"] → Range[NChannels]+L,
            ancillary,
            multiplexer,
            ancillary["Conjugate"]†,
            QuantumState[stateAnsatz, "Label" → "ψ(θ)"]["Dagger"] →
              Range[NChannels]+L
          }];
```

```
In[ ]:= ExpectedValueCircuit["Diagram"]
```

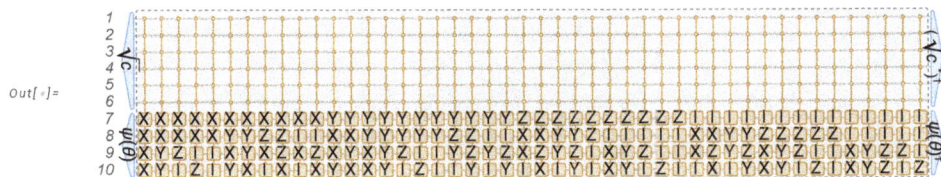

We can extract the energy expectation value using ComplexExpand:

```
In[ ]:= meanValue = ExpectedValueCircuit[]["Scalar"] // ComplexExpand;
       meanValue /. Thread[params → init] // Chop

       expectationValue[hamiltonian_QuantumOperator, state_QuantumState] :=
         QuantumCircuitOperator[{state, hamiltonian, state["Dagger"]}][]
       state = QuantumTensorProduct[
             Join[Table[QuantumState["0"], {i, 1, NChannels − 1}], {QuantumState["1"]}]];
       groundStateEnergy = expectationValue[Ham, state]["Amplitude"]〚1〛;
```

Energy Search

Cost Function

We can define the energy expectation value to be the cost function for the optimization algorithm. We will use the natural gradient descent:

```
In[ ]:= ClearAll[CostFunction];
       CostFunction[θ__] := meanValue /. Thread[params → {θ}]

In[ ]:= ClearAll[CostFunctionGrad];
       CostFunctionGrad[θ__] :=
             Grad[CostFunction[Sequence @@ params], params] /. Thread[params → {θ}];
```

Gradient Descent

To conduce a successful search of energy-minimizing parameters, we need to use the Fubini metric. The mapping between the parameter space of the ansatz and target Hilbert space is not isometric. We therefore should use the study metric. The gradient operator guides us downhill in the direction minimizing the energy. The Fubini study metric makes sure we walk carefully and don't take steps that are too large:

```
In[ ]:= metricTensor = FubiniStudyMetricTensor[stateAnsatz, params];
       ClearAll[fubini];
       fubini[θ__] := metricTensor /. Thread[params → {θ}];

In[ ]:= QuantumParameters = QuantumNaturalGradientDescent[CostFunction,
                 init, fubini[Sequence @@ #] &, "Jacobian" → CostFunctionGrad,
                 "LearningRate" → 0.1, "MaxIterations" → 200]; // AbsoluteTiming
```

In[ ]:= **ListLinePlot[Re[CostFunction @@@ QuantumParameters]]**

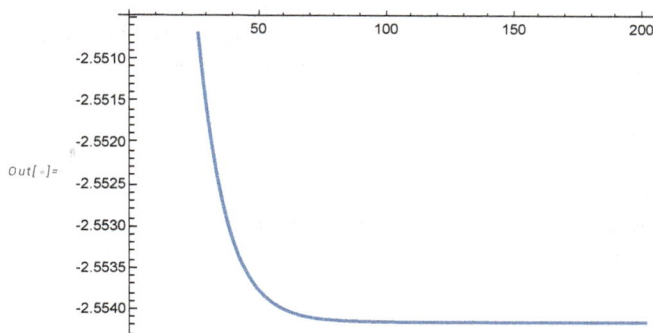

The resulting energy is quite far from the known value for helium. We can plot the amplitudes for different states:

In[ ]:= **AssociationThread[params → Last[QuantumParameters]];**
resultState = stateAnsatz[%];
resultState["ProbabilityPlot"]

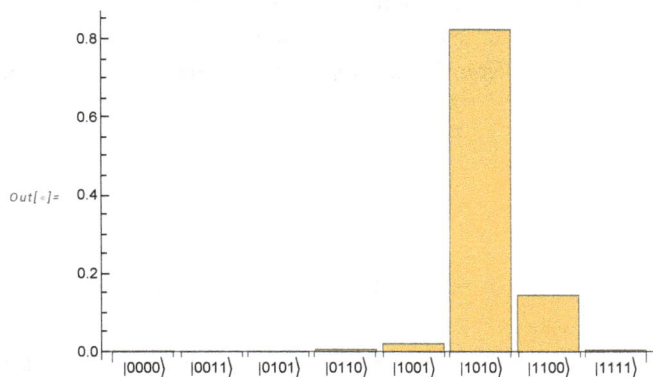

One may see that the state that minimizes the energy has a peak for the state with one of the electrons excited. It is nonsymmetric according to spin. This is not a problem in the case of our calculations, since states $|0101\rangle$ and $|1010\rangle$ are energetically indistinguishable. Our minimization algorithm converged to a region of the energy landscape where both states are degenerate, but the former state is disproportionately represented. It is also possible to observe that the energy discrepancy is rooted in the choice of the small computational basis of four qubits. Enlarging the basis would make the value of energy closer to the real value.

Concluding Remarks

The presented method is suitable for many-qubit, error-corrected quantum computers. It is not possible to use it for computation in its current state. There are also problems with the growing complexity of the circuit. Although we reduce the dimension of the parameter space from exponential to linear, we make a tradeoff with the quadratic growing complexity of the CNOT layer of the ansatz. Additionally, current mapping of the Hamiltonian grows exponentially in terms of circuit complexity. This growth can be suppressed by the truncation of Pauli gates with small weights. For atomic systems, however, it may not be sufficient because of the strong interactions of electron clouds and thus emerging strong correlations. The method is nonetheless suitable for quantum chemistry simulations where one can take into account valence band electrons distributed over many atoms.

The notebook can serve as a demonstration of the possibilities of the Wolfram Quantum Framework. In particular, it may serve as a guide on creating general classes of quantum circuits simulating atomic electron configurations.

Acknowledgments

I would like to thank my mentor Mads Bahrami for introducing me to the topic of quantum chemistry simulations using quantum circuits. I'd also like to thank my other mentor, Paul Abbott, for all the theoretical insight into helium atom simulations. I would like to thank my teaching assistants Sebastian Rodriguez and Bruno Tenorio for their priceless help in finding errors in my code and tunneling through all of the conceptual difficulties, as well as Payal Solanki for providing knowledge on how to construct a computational ansatz. Finally, I would like to thank Professor Stephen Wolfram for keeping an eye on my progress during the Wolfram Summer School.

References

1. H. A. Bethe and E. E. Salpeter (1957), *Quantum Mechanics of One- and Two-Electron Atoms.* Springer-Verlag.

2. P. C. Abbott and E. N. Malsen (1986), "A Model Wavefunction Including Electron Correlation for the Ground State of the Helium Isoelectronic Sequence," *Journal of Physics B: Atomic and Molecular Physics.* iopscience.iop.org/article/10.1088/0022-3700/19/11/014.

3. J. Tilly, et al. (2022), "The Variational Quantum Eigensolver: A Review of Methods and Best Practices," *Physics Reports* 986: 1–128. www.sciencedirect.com/science/article/pii /S0370157322003118.

4. J. D. Whitfield, et al. (2011), "Simulation of Electronic Structure Hamiltonians Using Quantum Computers," *Molecular Physics* 109(5): 735–750. www.tandfonline.com/doi/abs/10.1080 /00268976.2011.552441.

5. Wikipedia, "Jordan–Wigner transformation." en.wikipedia.org/wiki /Jordan%E2%80%93Wigner_transformation.

Access the Full Code

Scan or visit wolfr.am/WSS2024-Zdziennicki.

Cite This Notebook

"Electronic Structure of Atoms Using Quantum Computing Techniques"
by Michał Zdziennicki
Wolfram Community, STAFF PICKS, July 9, 2024
community.wolfram.com/groups/-/m/t/3209980

Building Blocks' Aggregation Systems

MARIA FERNANDA CASTRO ALVAREZ

Aggregation systems on grid graphs grow at each step by attaching a new atom to a preexisting configuration according to a set of local constraints. The multiway graphs of these systems fascinate us by sometimes filling space and other times reaching frustrated and dead surfaces. This project continues to explore aggregation systems on three-dimensional grid graphs by introducing unique and indivisible atoms, which adhere to specific combination rules. We examine configurations formed with a few special building blocks, whose interlocking mechanisms allow for practical experiments. These shapes, designed for the physical world, also have an abstract mathematical representation in terms of adjacency relations that form noteworthy graph structures. We introduce the concept of instruction graphs and ensure their uniqueness with regard to constructions in the

physical world. By comparing computational models to physical sculptures, we validate the practical applicability of our new approach, highlighting its potential to enhance our understanding and control of space-filling structures.

Introduction

Aggregation systems are built by adding new atoms to an already existing configuration. The purpose of this project is to explore all possible configurations using a given set of building blocks. The rules we will use to form our configurations are based on the physical characteristics of the building blocks: they will only bind together when their cuts and studs fit properly.

Our Approach

Building Blocks

In order to build a 3D aggregation system, we first need to choose a geometric shape to build upon. Starting with cubes, we construct different building blocks, each with specific "rules." We think of these building blocks as indivisible atoms that can only bind to another preexisting atom according to certain local constraints.

We use voxels to build each block and then use an association to distinguish where the "Ins" and "Outs" should be placed. Visually, "Ins" correspond to the cuts in each piece, while "Outs" are located on every stud. This association makes it possible to compute all the different configurations between the pieces and subsequently build an aggregation system with them.

All the pieces are built as a union of their cuts, main plate and studs. In the following image, we can see the process of building the L-shaped 2×2 plate:

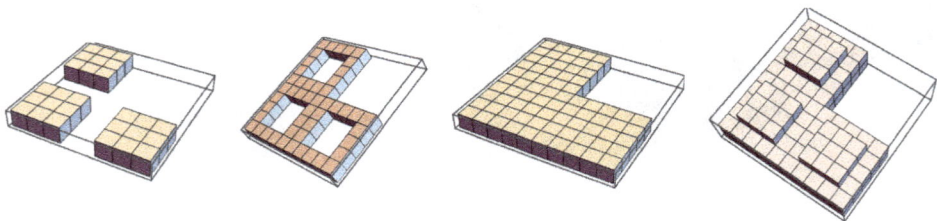

```
In[ ]:=  $BracketOPrototile = Association[
            "Name" → 99 781,
            "Voxels" →
              Union[cutsBracketO, plateBracketO, cubeBracketO, studsBracketO],
            "Ins" → Association[
                1 → {{4, 5, 4}, {0, 0, 1}},
                2 → {{9, 5, 4}, {0, 0, 1}}
              ],
```

```
     "Outs" → Association[
         3 → {{4, 5, 6}, {0, 0, 1}},
         4 → {{9, 5, 6}, {0, 0, 1}},
         5 → {{4, 1, 3}, {0, −1, 0}},
         6 → {{9, 1, 3}, {0, −1, 0}}
       ]
    ];
```

This specific example builds a bracket (99781) with two "Ins" and four "Outs". We repeat the same mechanism with all our desired pieces to obtain a set of pieces that will give our aggregation system more possible configurations. Part of the importance of having different pieces is also exploring their limitations: an aggregation system of 1×1 plates will allow a stack without ever encountering a dead or frustrated surface, but its adjacency graph will always be a path graph.

On the other hand, "Tiles" only contain "Ins", making it impossible to build an aggregation system without additional pieces. Their specific characteristics make them end pieces on every block, creating a dead surface wherever they're placed. The more pieces we add to our database, the larger the number of multiway systems and sculptures we will be able to build:

Rotations and Translations

Since we are building every single block with voxels, we only need to consider the ways a cube can rotate to account for all possible rotations of our pieces. A cube has 24 possible rotations, calculated as follows:

- The first rotation is the identity rotation, which means not doing anything.
- There are six rotations through the middle of one of the six pairs of opposing edges.
- There are nine rotations around the axes through the middle of one of the three pairs of opposing faces.
- Lastly, there are eight rotations through one of the four pairs of opposing corners.

Adding these up, we have $1 + 9 + 6 + 8 = 24$ possible rotations.

If we want to explore all possible configurations, we need not only a rotation function but also a translation function, so that we can stack blocks together.

Tiles

To display each building block, we use a function called InitializeTiling that includes not only the voxels but also the "Ins" and "Outs":

```
In[ ]:=  InitializeTiling[proto_] := Association[
            "Ins" → Values[proto["Ins"]],
            "Outs" → Values[proto["Outs"]],
            "Tiles" → Association[1 → proto],
            "Voxels" → proto["Voxels"]
         ]
```

Having a generalized function instead of calling each proto-tile individually each time is much easier and will be beneficial once we get to the multiway graphs and sculptures.

Multiway Graphs

Multiway graphs explore every possible configuration given a specific set of pieces. Our aggregation system adds one piece at a time, where every possibility is computed and explored. If we canonicalize them, we filter out any duplicates generated by rotation and symmetry. If not, the number of possibilities is much larger. Here's a clear comparison between a canonicalized multiway graph after one step and one without canonicalization of a 1×1 plate and a 1×2–1×4 bracket.

Multiway System (Canonicalized)

```
In[ ]:=  Module[{res, graph, block = $PlatePrototile},
            res = Reap[Nest[MultiIterate[
                          IterateBuilder[block],
                          CanonicalizeBy[
                             BuildingSignature,
                             ReverseSort]
                       ][#] &, { InitializeTiling[block]}, 1]];
            graph = SimpleGraph[Flatten[res[[2]]]];
            Graph[graph,
               AspectRatio → 1/3, EdgeStyle → Gray, VertexSize → 2,
               GraphLayout → "LayeredDigraphEmbedding",
               VertexSize → 1, PerformanceGoal → "Quality",
               VertexShapeFunction → Function[{off, val, scale},
                   Inset[DisplayBlocks[val], off, Automatic, First[scale]]
                  ]
              ]
            ]
```

Multiway System

```
In[ ]:=  Module[{res, graph, block = $PlatePrototile},
            res = Reap[Nest[MultiIterate[
                          IterateBuilder[block],
                          CanonicalizeBy[
                             Identity,
                             ReverseSort]
                       ][#] &, { InitializeTiling[block]}, 1]];
            graph = SimpleGraph[Flatten[res[[2]]]];
            Graph[graph,
               AspectRatio → 1/3, EdgeStyle → Gray, VertexSize → 2,
               GraphLayout → "LayeredDigraphEmbedding",
               VertexSize → 1, PerformanceGoal → "Quality",
               VertexShapeFunction → Function[{off, val, scale},
                   Inset[DisplayBlocks[val], off, Automatic, First[scale]]
                  ]
              ]
            ]
```

Out[ ]=

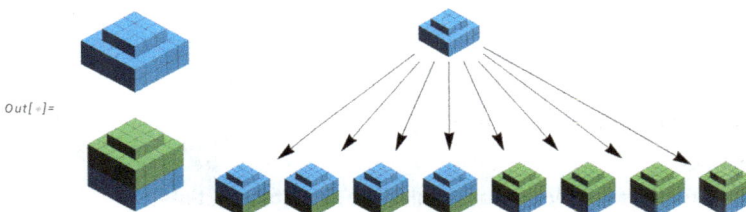

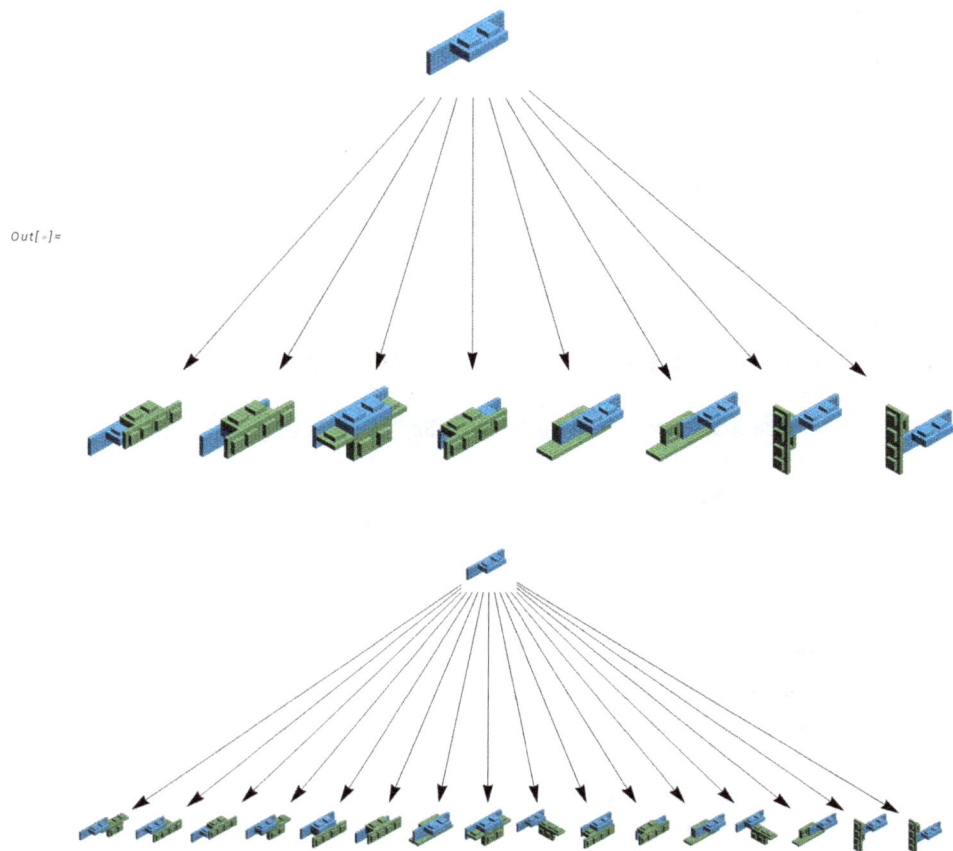

Out[·]=

Instruction Graphs and Sculptures

Sculptures

Each one of our so-called "sculptures" has been made using one of two methods: the Random Iterate Builder or the Explorer, which allows the user to dynamically choose the configuration of each piece (with the same physical restrictions that we would face in the real world).

The Random Iterate Builder chooses configurations with equal probability and keeps adding them one at a time. The only specifications needed are the piece we want to use for the sculpture and the number of pieces to be added.

Random Iterate Builder

```
In[ · ]:=  Module[{block = $BracketOPrototile, time = 50, rand},
          SeedRandom[123];
          rand = Nest[RandomIterateBuilder[block], InitializeTiling[block], time];
          Row[Show[#, ImageSize → 300] &/@
```

{DisplayBlocks[rand], BuildingAdjacencyGraph[rand]}]
]

Out[ ]=

Graphical Explorer

The Graphical Explorer allows us to work with more than one piece. The inputs to the function are all the desired pieces. It begins with an initialization piece that has to be specified. The "Ins" and "Outs" are marked with a black half-sphere resembling a button. Whenever the user clicks it, the Graphical Explorer will display all available pieces to continue. Once a new piece has been chosen, we click "Branch"; then it's time to pick which of the multiway paths we want to follow.

In[ ]:= HeadlightExplorer[{$BracketOPrototile, $Tile1x1Prototile},
    InitializeTiling[$BracketOPrototile]];

Out[ ]=

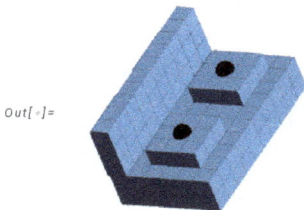

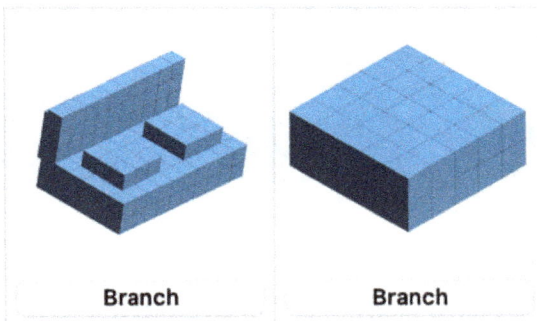

Branch   **Branch**

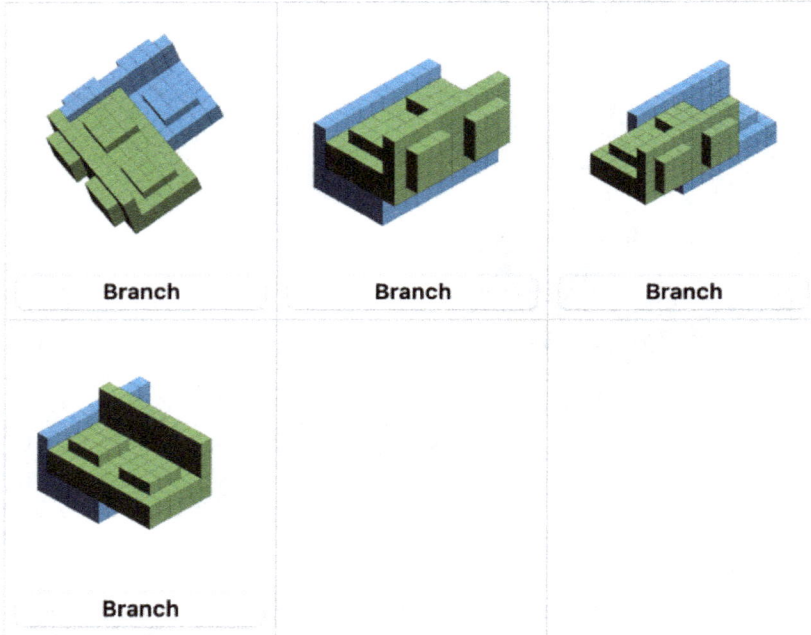

Branch	**Branch**	**Branch**

Branch

This process is repeated until we either encounter only dead surfaces or choose to stop the aggregation system voluntarily.

Instruction Graphs

Adjacency graphs can visually represent our building blocks' sculptures. Each vertex represents a piece, labeled as {i, "Name"} with i = 1, ..., n depending on the total number of pieces in the given "sculpture." If two vertices are connected by an edge, it indicates that at least one connection between their "Ins" and "Outs" has been made. Additionally, there is a label indicating the rotation of both pieces being joined:

In[ ]:= **deadSurfaceCube =** `‹...›` **;**

InstructionsGraph[deadSurfaceCube, ImageSize → {UpTo[300], UpTo[300]},
VertexLabels → Automatic, EdgeLabels → x_ :→ Last[x]]

Out[ ]=

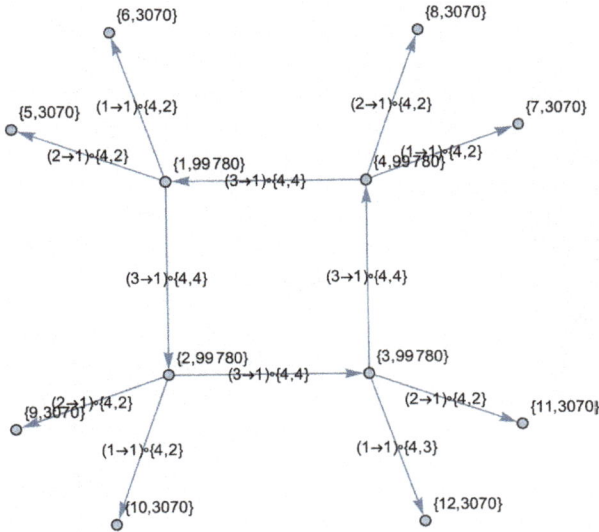

Mathematica already has an AdjacencyGraph function, but uniqueness isn't guaranteed, so we built our own. This is crucial because our goal is to construct an instruction graph based on an adjacency graph. To build a specific sculpture based on a graph, we need to know the rotation of each piece being added to the configuration as well as the specific "Ins" and "Outs". Once we incorporate this information into the graph, uniqueness can be assured, allowing us to build sculptures from the graph using a converter.

Converter

The GraphToBuilding function takes the graph's data as input, computes all the necessary information and outputs a sculpture with the specified specifications. We can test this with RandomIterateBuilder and observe that the only difference is the color, which is chosen randomly and has no importance beyond the aesthetic of the building:

```
In[•]:= Module[{block = $PartsDictionary[36 841], time = 8, rand, graph},
    SeedRandom[123];
    rand = Nest[RandomIterateBuilder[block], InitializeTiling[block], time];
    graph = InstructionsGraph[rand, ImageSize → {UpTo[300], UpTo[300]},
        VertexLabels → Automatic,  EdgeLabels → x_ :→ Last[x]];
    out = graph;
    {DisplayBlocks[rand], graph, DisplayBlocks[GraphToBuilding[graph]]}
    ]
```

RandomIterateBuilder graph:

BuildingAdjacencyGraph:

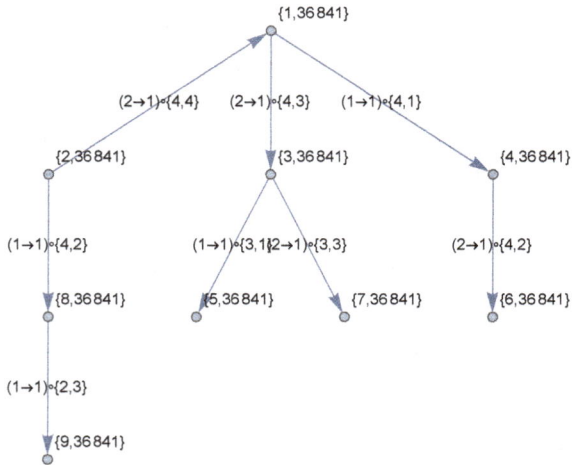

GraphToBuilding converted graph:

```
In[ ]:=  Module[{block = $PartsDictionary[111], time = 30, rand, graph},
         SeedRandom[123];
         rand = Nest[RandomIterateBuilder[block], InitializeTiling[block], time];
         graph = InstructionsGraph[rand, ImageSize → {UpTo[300], UpTo[300]},
             VertexLabels → Automatic, EdgeLabels → x_ :→ Last[x]];
         out = graph;
         {DisplayBlocks[rand], graph, DisplayBlocks[GraphToBuilding[graph]]}
         ]
```

RandomIterateBuilder graph:

BuildingAdjacencyGraph:

GraphToBuilding converted graph:

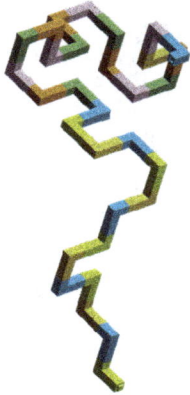

Other Sculptures and Their Graphs

All of these buildings were made using the Graphical Explorer and the InstructionsGraph function.

Closed-Path Graph

Pieces used:

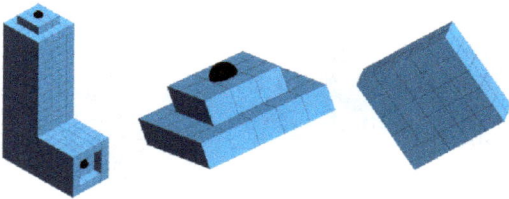

Sculpture:

InstructionsGraph:

Pinwheel

Pieces used:

Sculpture:

InstructionsGraph:

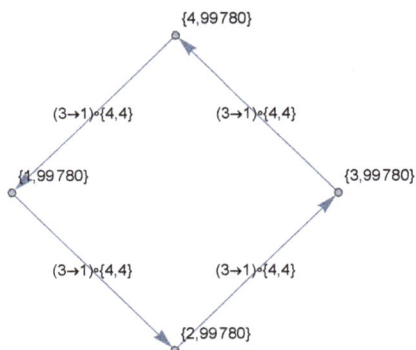

Dead Surface Ring

Pieces used:

Sculpture:

InstructionsGraph:

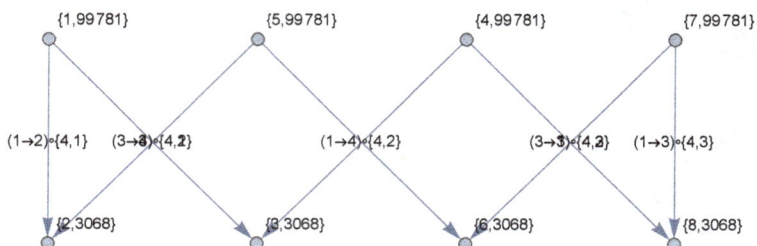

Concluding Remarks

Having conducted the analysis, it is intriguing to utilize both graphs and sculptures as inputs to generate 3D or graph representations. Employing building blocks for an aggregation system has granted us the freedom to establish our own rules, resulting in multiway systems that yield both artistic and analytical insights. Moving forward, expanding the range of pieces promises to further reveal the expansive potential of multiway systems.

Acknowledgements

I would like to thank everyone who had input in my project, but especially Brad Klee, who helped me not only with the project selection process but was also nothing but patient and helpful when it came to coding. This project builds upon "LEGO Meets Multiway Systems" by him, from which most of the functions used were found.

Reference

1. B. Klee (2024), "LEGO Meets Multiway Systems," *Wolfram Community*. community.wolfram.com/groups/-/m/t/3095175.

Access the Full Code

Scan or visit wolfr.am/WSS2024-Alvarez.

Cite This Notebook

"Building Blocks' Aggregation Systems"
by Maria Fernanda Castro Alvarez
Wolfram Community, STAFF PICKS, July 9, 2024
community.wolfram.com/groups/-/m/t/3210348

www.ingramcontent.com/pod-product-compliance
Lightning Source LLC
Chambersburg PA
CBHW051116200326
41518CB00016B/2523